Praise for *The Hungry Gene*:

"*The Hungry Gene* is a fair-minded, clear, and fascinating book about a highly emotional issue: why so many people are becoming so fat. After reading this book I have a much better sense of what science knows about the issue, what is unknown, what is a matter of individual destiny, and what is the result of food-industry strategies. An authoritative introduction to the next big public-health threat."

—James Fallows, national editor, *The Atlantic Monthly*

"Shell takes a provocative look at the history, politics, science and culture of obesity. . . . She is as adept at explaining the physiology of fat as she is at chronicling the desperation of the obese."

—*The Minneapolis Star-Tribune*

"Ellen Ruppel Shell's *The Hungry Gene* is important, richly informative, and written with flair."

—Alan P. Lightman, author of *Einstein's Dreams*

"Finally someone has done justice to what may be the most medically, socially, and commercially significant enterprise in all of science. An indefatigable reporter with a novelist's sense of character and drama, Ellen Ruppel Shell has poured all her considerable talent into this engrossing book."

—John Horgan, author of *The End of Science*

"In *The Hungry Gene*, Ellen Ruppel Shell gives us a clear, inviting, and entertaining look at this fascinating subject, and at the sociopolitical underpinnings to the mysterious—and frightening—obesity pandemic. Beautifully written and scrupulously researched, *The Hungry Gene* is for everyone who has worried—or wondered—about their weight."

—Jon Weiner, Pulitzer Prize–winning author of *The Beak of the Finch*

"Shell's book explores obesity both nationally and globally. She looks at it through the lens of history, science, economics and politics while interweaving its effects on individuals, cultures and countries. Most importantly, she addresses the nature-or-nurture question. It's both, says Shell. And it's complicated."

—Patricia Guthrie, *The Atlanta Journal-Constitution*

"Shell promotes the theory that the 'obesity pandemic' is a result of vulnerable genes in a fat-enabling culture. . . . The book goes beyond the earnest zeal, science and history of *Fast Food Nation* . . . to study the psychology and anthropology of fat as well. Once you've taken in the research from Micronesia to Maine and heard from the stomach-stapled, you may think the author's call for junk-food TV advertising is a good one. . . . It's also [a] relief to see a book hit the shelves this month that takes the issue far beyond the realm of 'miracle' diet guides that don't work."

—Leslie Yazel, *The Austin American-Statesman*

"Shell unleashes an all-out assault on consumer culture: as she sees it, an unholy trinity of television, cars, and fast food, all conspiring to ensure that we expend as little energy and consume as much bad food as possible."

—Joseph Hooper, *Elle*

"Shell's characters run the gamut from geneticists to nutritionists studying indigenous people of Micronesia, where a traditional diet of fish and breadfruit has been replaced by, of all things, Spam." —Stephanie Mencimer, *The Washington Monthly*

"This is a steel magnolia of a book. A political fist clenched in a scientific glove. Charmingly written with plenty of human interest and colour, its underlying message is clear and radical. The latest research in genetics and biochemistry is expertly presented, but this science isn't left hanging in a value moral limbo. All of the evidence points in the same direction." —*The Financial Times*

"Vividly conveys the excitement of scientific research conducted by not colourless rationalists, but by ingenious and persistent men of flesh and blood with a variety of motives, from scientific curiosity and altruism to thirst for glory and greed. Miss Shell deserves our gratitude for her timely, lucid and readable exposition." —*The Evening Standard*

"A fascinating study of the controversial hunt for the obesity gene."
—*The Herald* (Glasgow)

"*The Hungry Gene* has its quota of drama . . . it's all enthralling. [The author's] advice on what we can do about getting fat—and not getting fat—is well written and convincing."
—*New Scientist*

"A detective story . . . cleverly weaves science and historical fact . . . author brings the book alive with her investigative journalistic style." —*Nature*

"It is remarkable that a book as excellent as *The Hungry Gene* can be contained in such a slender volume. . . . [Shell] examines the history of obesity and suggests what can be done to control it. . . . As Shell points out, the obesity pandemic didn't begin the day McDonald's started selling fries." —Brad Evernson, *National Post*

"She looks at the issues surrounding our collective expansion. . . . Shell's disturbing look at obesity might just be the diet inspiration you need."
—Anne Stephenson, *The Arizona Republic*

"Obesity is quickly surpassing tobacco as the top threat to public health. . . . Shell investigates how Western-style commerce and culture have led to the fattening of people around the world and how scientists are isolating genetic, prenatal, and environmental factors that influence a person's body weight." —*Science News*

"This is the kind of book that you take everywhere, because you don't want to leave it, and then read out loud to friends and family and even casual acquaintances because it's so smart and so engaging. Ellen Ruppel Shell is a wonderful writer, and in *The Hungry Gene* she combines that ability with incisive science reporting to tell a story that—finally—makes some sense of our struggles with obesity."
—Deborah Blum, Pulitzer Prize–winning author of *Sex and the Brain*

The Hungry Gene

The Inside Story of the Obesity Industry

Ellen Ruppel Shell

Grove Press
New York

For Opa

Published simultaneously in Canada
Printed in Canada

FIRST GROVE PRESS EDITION

Library of Congress Cataloging-in-Publication Data
Shell, Ellen Ruppel, 1952–
The hungry gene : the inside story of the obesity industry / by Ellen Ruppel Shell.
p. cm.
Includes bibliographical references and index.
ISBN 0-8021-4033-5 (pbk.)
1. Obesity—Popular works. I. Title.
RC628 .S48 2002
616.3'98—dc21 2002071180

Grove Press
841 Broadway
New York, NY 10003

03 04 05 06 07 10 9 8 7 6 5 4 3 2 1

Contents

PREFACE
PAGE vii

INTRODUCTION: THE TRILLION-DOLLAR DISEASE
PAGE 1

1 A WEARINESS OF EATING
PAGE 6

2 WALKING NAKED
PAGE 23

3 NATURAL BORN FREAKS
PAGE 49

4 ON THE CUTTING EDGE
PAGE 66

5 HUNGER
PAGE 77

6 THE CLINICAL EXCEPTION
PAGE 105

7 COLLATERAL DAMAGE
PAGE 125

8 SPAMMED
PAGE 151

9 THE CHILD IS FATHER OF THE MAN
PAGE 173

10 AN ARM'S REACH FROM DESIRE
PAGE 191

11 THE RIGHT CHOICE
PAGE 220

NOTES
PAGE 237

ACKNOWLEDGMENTS
PAGE 279

INDEX
PAGE 285

Preface

While writing this book, I had no idea that obesity was on the cusp of becoming one of the most debated and discussed controversies of the early twenty-first century. Nor had I any idea that I was touching upon a national—and international—obsession. But I've since learned that obesity represents for millions the physical manifestation of a deep and abiding fear. So many of us have tried to control our appetites, and our weights, and so many of us have in the long run failed to do so. And the disappointment in this failure runs deep, far deeper than I had previously imagined.

Weight is widely considered a matter of personal responsibility, and on radio chat shows I've been chastised—sometimes bitterly—for suggesting otherwise. People call in to crow over how much weight they've lost on the latest diet—low fat, high fat, low carbohydrate, whatever—and to add that if others are too stupid or too lazy or too gluttonous to do the same, why should any of us care? After all, this is a free society, in which we make our own bloody choices, and if everyone would make the same choices as they've made, the world would be a thinner place. Rather than resorting to the obvious retort ("just wait: taking off weight is easy, keeping it off is not,"), I often tell a story of a young man I met in the course of researching this book, a boy who looked far older than his thirteen years. We were introduced in a hospital clinic, where he was being treated for Type II diabetes, an obesity-linked disorder once seen only in adults, but now increasingly common in children. I watched on the sidelines while he mounted the scale—carefully, one lumbering foot after the other—and the attendant slid the pointer just beyond the 300-pound mark. The boy looked impassive, resigned, and a little shamefaced as a nurse blitzed him with questions about his lifestyle: what did he eat, how much did he exercise, which I interpreted as code for: how did you manage to gain still more weight since your last appointment? Meanwhile, I spoke with his mother, a small woman, perhaps one third the weight of her son. She told me her boy ate only one meal a day, usually healthy food, like beans and rice. His weight problem, she said, was an absolute mystery; in particular given that his brothers were as slim as she. I asked her whether I might meet with her son privately, and she agreed. I waited while the boy plodded off to a nutritionist who lectured him once again on the importance of moderate portion size and regular exercise, and then we left, taking the elevator one floor down to the hospital lobby, where we all but fell into a fast food restaurant.

I now know that nearly 40 percent of the nation's hospitals have similar eating places, but at the time, the sight of such a place in the lobby of a hospital that specialized in treating children perplexed me. But I was desperate for a cup of coffee, and it was convenient, so I suggested we step inside. The boy lost his suspicious look, slouched through the doors, and up to the counter to order an early lunch. When the server asked the inevitable question, the boy didn't hesitate. Of course he wanted his burger, fries, and soda "supersized." Later he told me what he apparently hadn't told his mother—that fast food restaurants were a second home for him, a place he went before and after school, and with friends on weekends. He loved everything they had on offer, and especially the feeling of freedom it gave him, to make his own choices, to have it his way.

Television carries hourly reminders that having it "our way" is not a privilege, but a right, whenever and wherever we want it, even in the lobby of a clinic for treating children's obesity. It is increasingly difficult to resist, and indeed most of us don't, at least not for long. The odds of becoming overweight in much of the industrialized world today are suckers' odds—about three to one in favor of corpulence in the United States, two to one in the United Kingdom. The argument of personal responsibility weakens when one considers the consequences of this, which are enormous and borne by all of us. Industry is now beginning to recognize this: earlier this year a consortium of multinational corporations formed the Institute on the Costs and Health Effects of Obesity, focused on reducing obesity in the workplace and of course lowering the associated healthcare costs. To take but one example, Quebecor, a Canadian printing company, has calculated that fully 18 percent of the company's health benefits costs in 2003 will go toward employees with weight-related problems. Currently, about 9.4 percent of the national health care expenditures in the United States are directly related to obesity, but by the time you read this, that figure will almost certainly have edged upward. The obesity pandemic is by far the fastest growing public health crises in the industrialized world.

Naturally, the cost in human lives and health is horrifying, but what is motivating industry to change is fear of litigation and backlash, the threat to the bottom line. UBS, the international financial firm, recently warned of a "clear long-term risk to producers of fast foods, soft drinks, confectionery and snacks that anti-obesity measures will curb their ability to grow revenues in the future." In the United States, a trial attorney threatened to sue Kraft Foods for stuffing its products with unhealthy fat, and though he dropped the suit, Kraft nonetheless surprised the world by promising to stop marketing its products in schools. Another suit charging McDonald's with contributing to the obesity of children, though mocked by observers on both sides of the Atlantic, may in fact result in similar defensive actions.

Tackling this issue has not been easy, or comfortable. I've been put on industry "hit lists" and have been blasted by columnists in the business press. If anything, this has furthered my resolve, encouraged me to move forward. Resistant as vested interests are to change, they must, and they will. Indeed, it has already begun: in the course of just a few years, corporate involvement and concern over obesity has blossomed. Public awareness has peaked. This is all to the good, of course, but it is only the start of what is almost certain to become a serious and perhaps global movement. Such is the power of humans to change even the most entrenched institutions, and to challenge even the most firmly held assumptions. It is to this end that I hope this book will continue to contribute.

Ellen Ruppel Shell
July 2003

INTRODUCTION

THE TRILLION-DOLLAR DISEASE

Like most Americans of a certain age and sensibility, I am painfully conscious of body weight. I approach the bathroom scale with trepidation, swimsuit season with dread. But as a writer, I came to this issue obliquely, almost by chance. I write about science, and a few years ago I decided to learn something about the complex and forbidding subject I thought of vaguely as "genomics." For insight I went to speak with the chief business officer of one of the fastest-growing biotechnology companies in the world. He was a busy man and had penciled me in at an odd hour, after lunch on the day before Christmas, when the phones fell uncharacteristically silent.

At that time, this man was one of the canniest deal-makers in an industry of deal-makers. There was a public relations rep at his side and a lumpy sweatshirt emblazoned with his company logo on his back. He called the human genome the "book of life," and said grandly that now that the book had been decoded, his firm, and others like it, would "unpack the biology of disease" and transform "promiscuous" drugs into exquisitely guided missiles that would home in on and destroy specific disease targets. He said that this ordnance would be aimed at the big killers: cancer and heart disease.

As the man droned on, the day dimmed, and a slow wind picked up, hinting of snow. I began to worry that he might continue to talk straight through Christmas Eve. I had heard this rap before from other deal-makers, and I thought it disingenuous. Risking rudeness, I cut in to venture a question: If he were to focus his company's efforts on just one human affliction, what would it be?

The expected response—cancer or heart disease—was not the one he offered.

"Obesity," he said, eyes gleaming. "Obesity is the trillion-dollar disease."

I had not thought of obesity as a disease and I told him so.

"Three or four years ago, obesity was just a black box," he said. "Something having to do with gluttony, or a weakness of will. Now we know that will has very little to do with it. Now we know it's a problem that can be solved, just like any other."

The deal-maker had studied philosophy at Oxford, and was well versed in the nuances of Cartesian dualism, the ins and outs of the mind/body problem. It occurred to me that he might well be right. That a behavior as intimate and fundamental as eating was beyond human will, that appetite could, in a sense, have a "mind of its own" was intriguing, but also frightening. It posed opportunities—and dangers.

I had read books on the so-called biology of behavior, but until that moment I hadn't really believed them. In my heart I sided with Descartes—that the mind, the "thinking thing," was somehow separate from and nobler than the body, the "doing thing." Now I wondered. Obesity offered a striking example of a behavior—eating—that was not quite under willful control, of a machine that had somehow gained mastery over the mind. And for reasons no one quite understood, the machine was running amock.

Overweight is the most common and costly nutritional disorder of the twenty-first century, affecting 1.1 billion adults and a burgeoning number of children. In December 2001 U.S. Surgeon General David Satcher made public the casualties: 34 percent of adults overweight and an *additional* 27 percent obese. (A body mass index [BMI] of 30 is considered obese, a BMI of more than 25, overweight. This is calculated by dividing weight in kilograms by height in meters squared. A five-foot five-inch woman weighing 180 pounds has a BMI of 30, as does a six-foot two-inch man weighing 233 pounds.) Satcher estimated the costs attributed to overweight the previous year alone as $117 billion. Obesity, he warned, is quickly eclipsing tobacco as the number one threat to public health.

Obesity thrives on progress, and has hit hardest where Western-style commerce has taken root. Half the adult populations of Brazil, Chile, Colombia, Peru, Uruguay, Paraguay, England, Finland, and Russia are overweight or obese. The same goes for Bulgaria, Morocco, Mexico, and Saudi Arabia. In China, the world's most populous nation, obesity increased sixfold in the final decade of the twentieth century. In Japan, 20 percent of women and nearly one-quarter of men are overweight. In India, one of the world's poorest nations, overweight and obesity are endemic in the middle classes. And nowhere is the power of the disorder to change and mold a culture more evident than in the South Pacific, where on some islands as many as three-quarters of adults are dangerously obese.

Adults are not the only victims—far from it. In just two decades' time the incidence of overweight and obesity in American children has more than doubled. In Britain, youth obesity rates soared by 70 percent in a single decade. The trend holds true for Russia and China, Brazil and Australia. This is cause for alarm, for more than half of overweight children carry at least one risk factor for future cardiovascular disease; among these are high blood pressure, high blood cholesterol levels, and gyrating insulin levels. More frightening is that for the first time, obesity-linked "adult-onset" diabetes has trickled down into childhood, with one of every three new cases of the illness diagnosed in children. It is a sad irony that overnutrition has inflicted on youth a disorder whose symptoms—blindness, renal failure, and heart disease—mimic those normally associated with old age.

In her treatise, *Illness as Metaphor,* Susan Sontag wrote: "Any important disease whose causality is murky, and for which treatment is ineffectual, tends to be awash in significance." Obesity fits the bill on all counts: evoking for so many the deadly sins of sloth and gluttony, it elicits scorn as often as sympathy. Mythology shrouding the disorder has muffled sensible discourse and stymied research. For these and other reasons, obesity is among the most misunderstood of human conditions and among the most exploited. We have for decades argued without evidence over its etiology—often laying blame on the unwitting victims. The obese and overweight are mocked, shamed, and harassed; they are pummeled with bad advice and patronized with marketing campaigns. Only the marketing has paid off.

Americans spend $33 billion a year—more than the gross national product of most developing countries—on weight loss products and schemes: for frozen confections with the metallic tang of sugar substitute, fat-free snacks with the gastrointestinal impact of mineral oil, body-reducing contraptions, diet workshops, and for-profit support groups. We sign lifetime contracts with health clubs, hire personal trainers, buy diet books and low-fat foods by the ton, and have all but canonized Oprah Winfrey for her sporadic triumphs over the powerful pull of adiposity. We spend hundreds of millions on weight loss medications that can sicken or even kill us, and risk death on the operating table. And all the while, we grow fatter.

"Obesity science" has, in the last half-decade, been transformed from a lackluster backwater into a vibrant field of inquiry attracting some of the most brilliant minds in science, and liberating overweight from the murky ghetto of "character flaw" to the more potent status of "disease." Startling new insights into the genetic, prenatal, and environmental factors directing body weight have opened deeper questions and bigger challenges. They also point temptingly to concrete and viable approaches to prevention and treatment.

Science, though critical, is only part of the tale, setting the groundwork for what is to come. The food and drug industries play important roles, as do the lobbyists and public servants who inhabit—wittingly or not—the sociopolitical underbelly of the obesity pandemic. Along the way are the social scientists, historians, and public health experts who have illuminated and refined our understanding of obesity and recast it in a new and more hopeful perspective. It is among these that the heroes emerge.

This book, then, is about how the world got fat and what we can do about it.

The story of obesity is a sometimes twisted, sometimes wacky tale that sheds light on the best and the worst in all of us. There is intrigue here, and melodrama; a quest for knowledge, certainly, but also for intellectual primacy, glory, power, and wealth. At the heart of the story are the people whose lives are bound by their struggle with weight. It is their stories—the eminently human ones—that are the common thread, and it is their lives that this book is meant to change. Progress and prosperity have brought us the means to subvert the delicate balance the human body achieved through hundreds of millennia of evolution. Science has given us the understanding—and the means—to regain that balance. What follows is the challenge that we do so.

ONE

A WEARINESS OF EATING

If you examine a man who suffers from his stomach. All his limbs are heavy. You find his stomach is dragging. It goes and comes under your fingers. Then you shall say concerning him: this is a weariness of eating.

—*The Egyptian Book of the Stomach*

The first time I set eyes on Nancy Wright, she is flat on her back and cruciate. She is vaguely pretty, her eyes frightened but oddly beguiling. Her thick hair is loose and wavy, auburn with a sly touch of gray at the temples. You can see what some men see in her, and also, perhaps, why two husbands have come and gone. Even as she lies splayed and sedated on a gurney in Operating Room 17 at Beth Israel Deaconess Medical Center in Boston, you can sense that Nancy Wright is possessed of an immutable will.

Nancy once told me that she'd started out life large and kept on going. She didn't mean it as a joke. She weighed ten pounds, four ounces when she came into this world, and through childhood ate herself so big that her father thought she had psychological problems. Nancy didn't see it that way, but she did know that her relationship with food was tempestuous, like a doomed love affair. "Food has always been my best friend and worst enemy rolled into one," she told me. Now, in middle age, this dysfunctional relationship has made even simple pleasures difficult. It is getting harder for her to work in her flower garden, harder to play with her five grandchildren. And she keeps getting sick. She has hypertension, high blood cholesterol, and sleep apnea. She hates being so tired all the time, and so feeble, and she has done everything she can think of to fight it. She has tried Weight Watchers, Jenny Craig, and diet pills. All of these worked, for a while. The pounds melted away, and Nancy thought she'd found salvation. She'd buy new clothes and start making plans for a new life. But then, without knowing why, she'd fall off the wagon, and her old life would rush back. It was like waking up to a nightmare.

People tell Nancy she lacks willpower, but they are wrong. She has plenty. She stayed with the same thankless social services job for twenty years. She stayed with the same thankless husband for nineteen. And as a fiftieth birthday present to herself, she quit smoking. She hasn't touched a cigarette in four years, and doesn't plan to touch one ever again. But food is another matter. "You can live without cigarettes," she said, "but you have to eat."

It all comes down to a balance of power—or, rather, to an imbalance. Nancy can no more tame her compulsion to eat than a marooned sailor can tame his thirst. For Nancy, food is more than an addiction, it is like breathing—a constant, throbbing need.

Dr. Edward Mun understands all this perfectly. Mun is an assistant professor of surgery at Harvard Medical School, and an attending surgeon at Beth Israel Deaconess Medical Center. At thirty-eight he has the self-assured manner and polished good looks of a man born to take charge. But beneath the Ivy League veneer and the designer suit lie hints of a nerdy immigrant boy, a gawky overeager kid who spent his summers squinting through a microscope at science camp rather than hanging out at Little League with his pals. Like Nancy, Ed Mun hasn't always fit in. He was born in Korea and grew up in Gardena, California, the son of restaurant owners who expected much more of their boy than they themselves had managed to achieve. Young Ed did not disappoint; he was the model of the good Asian son. He aced high school, enrolled in Yale University, and graduated in four years with both bachelor's and master's degrees in biochemistry. He further distinguished himself at Harvard Medical School, and nabbed a coveted surgical residency in sunny San Diego. Surgery offered the most money, the most prestige, and the greatest opportunity to perform technically interesting procedures. But Mun wanted more.

"I wanted neurosurgery because I thought it was only for the talented few," he said. "But the truth is that there aren't that many brain operations. Neurosurgeons do herniated disks and trauma cases. Mostly, it's boring."

So Mun returned to Boston, to Harvard, and to Beth Israel Deaconess Hospital, to apprentice in general surgery. He removed breast cancers and performed stomach surgery. To his great relief, he didn't find this boring at all. But he did find it frustrating. Breast cancer patients had the habit of scrutinizing the Internet for facts about their disease and hauling reams of downloaded information to his office for review. Mun didn't like the messiness of that, the presumptuousness. Breast cancer, he says, is usually a matter of small incisions and quick recoveries. Yet the patients would piss and moan and demand second opinions. He didn't mind the second opinions, of course, but he did mind being put through the third degree. And he blanched at their sense of entitlement. These women were hot reactors, the sort of patients who required more assurance than he had to offer.

But the stomach was something else altogether. He liked the feel of it, the hard muscularity of the thing. And he liked that stomach patients trusted him, put themselves in his hands. They didn't ask a lot of extraneous questions, didn't expect miracles. He found stomach surgery enthralling, so much so that one would think he had some sort of a belly fetish. But it was nothing like that.

"In Japan and in Korea, tens of thousands of people die from stomach cancer every year," he said. "I also lost several relatives to this disease."

In Korea stomach surgeons are held in the highest esteem. Among these masters was Mun's paternal grandfather, the man in whose steps young Ed was meant to follow. Mun very much wanted to be like his grandfather, and to earn the respect of his demanding parents. So he studied and worked until he became one of the best stomach surgeons at Beth Israel Deaconess Medical Center, which is to say one of the best in the country, and perhaps in the world. But unlike his grandfather, Mun doesn't open many bellies to remove cancerous lesions or to repair ulcers. What Mun does mostly is something very few Korean surgeons—and only a few American surgeons—have ever done or would ever dream of doing. What Mun does is to take perfectly healthy stomachs and replumb them, cutting them loose from their natural moorings at the end of the esophagus and fashioning them into pouches the size of robin's eggs. This procedure, which generally takes Mun about ninety minutes but most other surgeons much longer, is called a Roux-en-Y gastric bypass.

Stomach surgery is a pretty rough ride. People who get their guts whittled and rearranged in this way can't eat much for weeks afterwards, certainly not nearly as much as they did before the change. If for some reason they succumb to the temptation to eat more than the little that their stomach can hold, they vomit. Vomiting is not really a complication of gastric bypass surgery; it is an expected and important side effect.

Gastric bypass patients sometimes lose so much weight that old friends and relatives barely recognize them. The surgery is reserved for those with one hundred or more pounds to lose. On average, patients shed about 60 percent of their excess weight in about eighteen months. It is hard to imagine many people in Korea being interested in such an operation. But in 2000, the year I met Mun, forty thousand Americans underwent gastric bypass surgery, about double the number performed only five years earlier. That number was expected to nearly double again by 2003. Mun doesn't find these figures the least surprising; he knows many people require his services. In Boston his dance card is full. And Nancy is next in line.

* * *

Nancy is five feet three inches tall and, at the time of her operation, weighs 274 pounds. Her BMI is 48.5, well into the morbidly obese range and she thinks she would feel and look much better if she were one hundred pounds or more lighter. She has seen what gastric bypass surgery can do for people: two of her coworkers have been transformed by the procedure, and a year ago her youngest daughter underwent the surgery and dropped ninety pounds. It was her daughter especially who convinced Nancy to give surgery a go, not so much with her words, but by her example. Both mother and daughter, Nancy said, are stubborn as mules. She figures that if gastric bypass worked for her daughter, it will work just as well for her.

Mun doesn't know Nancy is stubborn, but he does know that she is an especially good candidate for obesity surgery. For one thing, she is in relatively good health, without the horrific complications suffered by so many of the morbidly obese. For another, she is relatively small. A man of Mun's experience might well see Nancy that way. The hospital bed waiting outside his operating room is a "Big Boy," built to hold up to five hundred pounds. Sometimes it takes two Big Boys pushed side by side to hold one of Mun's gastric bypass patients. Mun remembers a seven-hundred-pounder whom he envisioned falling on him and crushing him to death. Nancy evokes no such grim images. There will be room to spare on her Big Boy.

Still, Mun has not promised Nancy success, or even survival. Gastric bypass kills one out of a hundred patients on the operating table, and not everyone recovers from its complications. There are few controlled studies of the procedure, so no one can speak with authority on its degree of danger. Still, the insurance industry classifies it as "high risk." The anesthesiologist on duty warns that corpulent patients are tricky, and that Nancy is no exception. Nancy's veins are buried in a thick layer of fat, making it hard for a needle to find purchase. Like many obese people, her tongue is large and her neck short, making it difficult to guide a breathing tube down her windpipe. It takes a bevy of nurses and doctors several attempts to finagle each of these maneuvers, and with every attempt it looks like Nancy will choke or cry. But she doesn't, and with time and effort the requisite tubes and needles get coaxed and jabbed into place. Nancy's eyes flutter and close and the anesthesiologist tapes them shut to prevent the corneas from drying out. Paralyzed from the anesthesia, Nancy draws her breath by machine. Plastic shrouds her face, presumably to shield wayward gore. She emanates fewer signs of life than do the machines to which she is tethered.

Mun helps the nurses arrange the layers of sterile drape, leaving exposed a rectangle of stark white skin roughly the area of a shoe box lid, size ten. He paints the rectangle orange with antiseptic. The flesh ripples thickly, like a crème brûlée. Mun grabs a black ballpoint pen and traces down the center line, a little shaky at first, then more or less finding the line he is after, about an eight-inch stretch from the tip of the breastbone to the navel. Seble Gabre-Madhin, a surgical resident, accepts a cauterizing scalpel from a nurse and traces over that line again and again until the skin bursts open with the force of the fat beneath. An observing medical student startles. It's not the sight that makes him queasy, he whispers, it is the smell, which is savory, like hamburgers spitting on a grill. The translucent fat layer glistens yellow under the operating room lights. The attending nurses hover. Drs. Gabre-Madhin and Mun exchange looks, then press two palms each on either side of the neatly split skin and ease the fat apart, forming a canyon. The walls of the canyon are slippery and lightly variegated with red blood vessels. There is almost no blood.

In *The Wisdom of the Body,* Sherwin B. Nuland, a clinical professor of surgery at Yale University, writes that the stomach is best understood "seen as a large bag near the upper end of what is otherwise a hollow muscular tube some twenty-five feet long from mouth to anus, the central portion of which is coiled up in the abdomen." This tube is the gut, and from stem to stern it comprises the pharynx, the esophagus, the stomach, the small intestine, the large intestine (or colon), and the rectum. The gut has an inner and outer layer of muscle, and the stomach has yet a third, to aid in its tireless churning of food.

The muscles and fibrous layers covering both sides of the stomach wall meet and fuse together in the middle of the belly, forming the linea alba, a stout ribbon of tissue stretching from the breastbone to the pubis. Mun deftly splits this, exposing the well-packed contents of the abdominal cavity, the largest orifice in the human body. Nurses position a gray metal circular retractor to hold back the skin and flab. The crater yawns jagged and raw. Mun pulls a glutinous apron of fat and blood vessels outside the wound, and lays it to one side of the torso. The mess on the surgical sheet is ghastly, like a mangled tongue lolling from the mouth of a drunk.

Mun plays archaeologist, pointing out artifacts as he excavates. Plunging his hand into the cavity, he locates the expected umbilical hernia, a weakness in the muscle near the belly button that is common in the obese. Wrist deep, he palpates the taut purple liver. He had mentioned earlier that the livers of the obese can grow monstrous—"sometimes," he told me, "they are as big as a horse's liver." This liver, thank goodness, is not Clydesdale-sized. Mun gently retracts it to examine the junction between the stomach and the esophagus. He is now elbow deep, pawing blind for the start of the stomach. Long seconds pass, and Mun's brow arches in concentration. No one says a word. This is a tricky business, and even the assisting surgeon, a stout, world-weary young woman, seems to hold her breath.

Suddenly, Mun finds what he's after. He stops for a moment and looks back at me, triumphant in his sterile mask and lightly fogged glasses.

"I love this organ," he says, pulling the stomach into glorious view.

Bariatric surgery, as obesity surgery is called, has a controversial history dating back hundreds of years. But the first modern procedure on record was in 1889, performed by Howard A. Kelly, a founding member of the faculty at Johns Hopkins University and its first professor of obstetrics. Kelly was an inventive surgeon and developed numerous surgical devices as well as innovative operative procedures. He seems to have fancied himself quite a sculptor, for he carved layers of fat from the abdomens of unwitting patients while they were under the knife for other problems.

Over the next few decades, reports of similar adventures trickled in from France, Germany, and Russia, and by the early 1920s, obesity surgery had become, if not fashionable, at least less ignominious. Not all obesity surgery patients died from massive infection or blood loss, but enough did that eventually it became clear to most respectable surgeons that slicing large quantities of fat from human bodies was not necessarily the safe and sure approach to treating the overweight that enthusiasts had claimed.

By the mid-twentieth century, obesity surgery had fallen out of favor with all but a few die-hard zealots. George Blackburn was not one of these. Blackburn is a surgeon, and is now director of the Center for the Study of Nutrition and Medicine at Beth Israel Deaconess Medical Center, where Edward Mun works. He is of medium height and robust, with stark white hair cropped into a schoolboy fringe. When we meet he radiates the sort of deep, smoky tan that comes from riding shotgun on a golf cart. There is no telling whether Blackburn plays golf, but I later learn that he gave up performing surgery years ago. He remains a darling on the obesity circuit, however, lecturing at conferences, consulting with industry, and sitting on boards and committees.

Like Mun, Blackburn trained at Harvard as a general surgeon. Like Mun, as a young man he developed a special interest in the stomach and in disorders of the gastrointestinal tract, such as ulcers. He cultivated this interest in the early 1960s, when doctors were not yet aware that ulcers are frequently caused by bacteria and treatable with antibiotics. Back then, dietary changes were the most common ulcer palliative, followed by surgical treatment. Surgery for severe ulcers usually involved removal of part or all of the stomach. It is possible to live and even to thrive without a stomach by eating many small meals and taking daily vitamin injections. Still, a hefty percentage of ulcer patients died of postoperative bleeding, infection, and sometimes starvation. Starvation was also a problem for trauma patients and for patients who lost their appetites after surgery. Blackburn got interested in this problem and decided to make it a subject of study in the early 1970s. To study it, he needed to experiment, and to experiment, he needed people who were willing to starve. He put ads in the Boston newspapers, assuming that he would get little if any response. "I was overwhelmed by the number of people who volunteered," he says. "Stunned."

Even more stunning than the number of volunteers was their size. Most were overweight or obese. Blackburn had very little experience with obese patients, and he was unfamiliar with their ways. He assumed, as did most doctors at the time, that fat people were too gluttonous to forgo a single meal, let alone subject themselves to weeks without food. But the overweight and obese volunteers were more than happy to starve. And they didn't cheat. They faithfully followed Blackburn's orders, eating only very small amounts of fish, fowl, or meat to keep up their protein levels to maintain as much muscle mass as possible. A month and a half later, they were many pounds lighter, and surprisingly healthy. Thanks to the "protein sparing" regimen, they showed no signs of the muscle wasting or dehydration usually associated with starvation regimens. That was all Blackburn needed to know, and he thanked the volunteers and told them it was time to leave. But he recalls that many—maybe most—begged to stay. "They would do anything to lose weight," he says. "And I mean anything."

Blackburn was a surgeon, not an endocrinologist, and he saw the plight of these patients through a surgeon's eyes. He was intrigued and moved by their pleas, but he had observed obesity surgery as a resident, and he hadn't liked what he'd seen. True, it had gone beyond the nineteenth-century "slice and dice," but to his way of thinking it hadn't gone far enough. The technique of choice at that time was the jejunoileal bypass, in which the intestine was essentially short-circuited to allow most of what the patients ate to slip through unnoticed and unabsorbed by the gut. This method was fairly effective, but the side effects—infection, protein malnutrition, kidney stones, osteoporosis, anemia, and liver failure—were unpredictable and occasionally fatal. "Some people would rather die than be fat," Blackburn says. And some doctors colluded in the gamble. About 100,000 intestinal bypasses were performed in the 1960s and early 1970s. Still, the high failure and mortality rates of intestinal bypass were troubling, and Blackburn chose to stay clear of the procedure.

There were other more benign approaches to consider—jaw wiring, for example. Jaw wiring is exactly as it sounds, something Wile E. Coyote might cook up for the Road Runner. As described in the British medical journal, *The Lancet*, the wiring procedure was simple enough: "Two interdental eyelets were placed in each canine and pre-molar region under local anaesthetic and the eyelets on opposing jaws wired together. Instruction was given on oral hygiene, measures to avoid aspiration, and the use of wire cutters." These "instructions" notwithstanding, expressing oneself through bound jaws was a trial, as was brushing one's teeth or eating anything that could not be slurped through a straw. And vomiting while wired could be lethal. (That is, if one somehow forgot to bring one's wire cutters to an office picnic.) Accidents—and deaths—did occur. And those who endured the recommended six months were horrified to find their weight soar when their jaws—and their appetites—were unleashed. Some physicians tried to stave off this rebound by prescribing a waist cord—a nylon strap tied tightly around the middle of slimmed patients that would remind them to eat sparingly. But most patients did not allow mere nylon to come between them and their calories, and either cut the dreaded things off or allowed themselves to balloon into giant hourglasses, with the waist cord strangling their middles like a noose. Clearly, when it came to weight loss, bondage was not the answer.

A more elaborate and certainly more imaginative scheme was the intragastric balloon, threaded into the stomach and blown up to crowd out the space for food. Stomach balloons carried a sort of whimsical allure: what could be more benign than a whisper of latex pumped with air? But the procedure was surprisingly expensive and not particularly effective. It was also dangerous, causing ulcers and erosion of the stomach lining. In 1988 a prominent physician decried the practice as "balloonacy" in an article in the journal *Gastroenterology.* Surgeons of the time were also experimenting with esophageal banding—inspired, perhaps, by the picturesque cormorant fishers of Japan. Cormorant fishing, or *ukai,* involves binding the necks of tame cormorants and setting them free to scoop sweetfish from the Nagara River. The birds are then called back or hauled in by their keepers and the fish extracted from halfway down their gullets. Esophageal banding has worked well for generations of fishermen in Japan, where the sport has become something of a tourist attraction. But it enjoyed only a short run in humans, who developed severe and sometimes fatal infections of the esophagus from the binding, and in any case had a greater tendency than birds to complain about fish getting stuck in their throats.

These and other disappointments prompted obesity surgeons to reconsider their options. Should they perhaps return to the stomach as a primary target? After all, the stomach's day job is churning great wads of stuff in a vat of acid, so it is accustomed to rough treatment and less prone to injury than are other more delicate organs. Also, at least theoretically, shrinking the stomach directly limited the amount of space available for food. And for many people a smaller stomach demanded less food than did a larger one. As explained by Columbia University College of Physicians & Surgeons professor Michael D. Gershon, author of an ode to the digestive tract, *The Second Brain,* when the stomach is full, receptors that respond both to pressure and to nutrients put a signaling system in motion to stop eating. A smaller stomach brings this pressure to bear more quickly. By shrinking the stomach, surgeons discovered that they could curb not only the intake of food, but also, in a surprising number of cases, the desire for food.

Stomach shrinking may not be brain surgery, but it has its challenges. Early approaches involved roping the stomach off near its throat with a one-centimeter-wide plastic tourniquet. There was no cutting, no stitches, no piercing of the stomach wall. The downside was that the bands sometimes worked their way into the stomach's interior, and got tangled into a mess that adhered to the stomach lining, causing pain and damage. A somewhat more complicated version of this approach, vertical-banded gastroplasty, eliminated the tangling problem and became quite popular. Both methods remain in use today, but at Beth Israel Deaconess Medical Center, George Blackburn does not recommend them.

In his drab and sprawling first-floor office, Blackburn keeps what he calls a "sympathy suit," a sort of hair shirt he pulls out for visiting doctors and scientists. The suit is shiny and puffy, like astronaut gear, and is filled with sand. It weighs thirty-four pounds, and donning it is meant to give skeptics some idea of how obese people feel every waking minute. Blackburn believes that the risk of obesity surgery and its complications is nothing compared with the risk of dragging all that excess baggage through life. Although he would almost certainly object, it would not be a stretch to call Blackburn a gastric bypass evangelist. He has seen what obesity can do to the human spirit, and it horrifies him.

"When I see a young man, a football player type of, say, 280 pounds, I'm able to look into the crystal ball of his future and see diabetes, insulin resistance, a whole slew of problems. Who among us would intentionally let people walk into the lion's mouth of obesity disease? No, I don't consider surgery drastic in the least."

Blackburn directs me to Isaac Greenberg, a psychologist who has worked with him for years. Greenberg is on the tall side and slim as a greyhound, but he's not smug about this. "Thin people think they're thin because they are doing something right," he says. "But they are wrong. What surgery has done is blow the psychoanalytic theories of obesity to hell. After the operation, many people lose their obsession with food. No one knows why, but it certainly proves that obesity is not just a psychological disorder."

Greenberg screens patients to see whether they are suitable candidates for gastric bypass. Most of these people have tried every weight loss scheme in the book. "Programs like Jenny Craig or Weight Watchers are like a virus," Greenberg says. "People go to meetings, see that other people have lost weight, and they want to catch that virus, too. So they sign up. They lose weight, go off the program, go off the special foods and gain the weight back. So, they return to the program. There are millions of returning customers keeping those outfits in business."

Greenberg invites me to attend a monthly obesity surgery support group meeting, and one dreary Thursday evening I decide to take him up on it. The group gathers in a basement auditorium at the hospital at dinnertime, but food is not served. The basement is dingy and echoing, and furnished with gray folding chairs. A ramp allows people too large to walk to roll into the room in wheelchairs. Four chairs roll in this evening. In all there are about three dozen participants. Most are in their thirties and forties, but they look older, their faces pasty and drawn, their bodies bloated with disuse. Their thick legs look vestigial. They complain of heart disease, diabetes, arthritis, and their difficulties in finding comfortable shoes. They are roughly divided between those who are scheduled for surgery and those who have recently had it. Few of the postoperative patients are anything close to thin. They have complaints—vomiting, getting food stuck halfway down their throats, hernias, gallstones—and they look just as tired, saggy, and worn as the people who have not yet had the operation. But most seem to agree that these discomforts pale compared to the agony they had endured throughout their presurgical lives.

It is no secret that the very obese live lives of quiet desperation, of humiliation and isolation, and of relentless guilt. Children have said in surveys that they would prefer playmates with missing legs or eyes to those with too much fat. Many obese adults confess that they would prefer to be blind or deaf. No wonder, then, that the fat people who gather together in this dank auditorium would, as George Blackburn predicted, do anything to be thin.

Among them only two, a media manager and a computer programmer, show any visible effect of the surgery. The programmer has lost one hundred pounds. He has 150 pounds still to lose, and he looks peaked, as though fighting a losing bout with the flu. The manager, a tall woman in her early thirties, announces that she is fourteen weeks pregnant. She looks terrific.

"One hundred thirty-seven pounds ago, it took nearly two years for me to get pregnant with my first child," she says. "This time it took one try." The room shakes with applause. After the cheers die down, a support group member raises a timid hand to ask if there is "any downside" to the surgery. There certainly is, the manager says cheerfully. "For weeks after the operation I thought I'd made a mistake. I was so sick I could hardly move. I really did think I was going to die."

Nancy Wright has attended these support group meetings, and she knows exactly what she is in for. She knows that dying on the operating table is a distinct possibility. She knows about the complications—the exhausting anemia, the painful gallstones and incisional hernias, the infections. Hearing about the ghastly aftereffects of the surgery gave her pause, but not doubt. She hasn't actually witnessed a gastric bypass, doesn't know in explicit detail all it entails, but she has spoken with Dr. Mun, and she has perfect faith in him.

And Ed Mun has perfect faith in himself. He has completed 110 gastric bypass procedures in the past twelve months, and most have gone like clockwork. One patient blew out her staple line postoperatively and wound up in the intensive care unit for thirty-five days. But she'd had three kidney transplants and a suppressed immune system and was a bad candidate for the procedure from the get-go. Nancy is a textbook case.

Mun reminds me of this as he manipulates Nancy's guts, maneuvering the left side of her stomach down into the operating field and into his line of vision. He takes care not to damage the spleen, which bleeds uncontrollably if torn, or the pancreas, which is filled with enzymes that can leak out and, as he puts it, "eat everything up." The stomach is glistening, firm. It is easy to imagine how demanding that stomach can be. Mun sizes it up and gets to work. Taking aim with his Endo-GIA II stapler, which cuts as well as rivets, he divides the stomach into two parts and staples the top portion into a twenty-milliliter pouch, about two tablespoons. An organ that could once proudly contain a quart of Häagen-Dasz can now barely hold a shot glass of yogurt. The bottom section will still secrete stomach juice, but it won't nag Nancy for food. Mun feels down the small intestine, past the duodenum to the jejunum, slices through the sausagelike structure and nails it to the stomach pouch with his EEA circular stapler, creating a passageway the circumference of an M&M. Mun regards the EEA fondly, and praises its smooth, powerful action. The device spits out sixty-four stainless steel staples at a trigger pull, accomplishing in a split second what once took long minutes of painstaking needlework. Like a rodeo star roping a calf, Mun grabs the slippery bypassed end of the small intestine, flips it, and staples it to the lower small intestine, forming a Y. Stapling the jejunum to the stomach pouch will keep Nancy from cheating. For some reason doctors can't quite agree on, contorting the gut this way causes it to shoot caution signals to the brain when fatty or, in particular, sweet food passes through. These signals spark a violent "dumping" reaction—the body rejects the treat as it would poison, sparking cramps, hot flashes, pain, and diarrhea. For a few months at least, Nancy's stomach will no longer have the power to demand a Snickers bar. Mun has tamed it.

Looking up, Mun checks his watch: it is just over an hour since the first incision. Inhaling deeply, he stands back and admires his work: perfect, and in less time than it takes most people to balance their checkbooks.

Three hours after the operation, Nancy is awake and hurting in her Big Boy. She is exhausted, gray with fatigue and pain. "That was no walk in the park," she says. I ask if she has any regrets, or—worse yet—any thoughts of food. "No way," she says, her eyelids fluttering. "That's behind me now."

At nine o'clock the next morning, Edward Mun performs his seventh gastric bypass of the week. He is to leave that night for Korea, to attend his elementary school reunion. The fact that he has agreed to fly halfway around the globe to gather with childhood friends somehow doesn't surprise me. Mun is a man of deep loyalty and total dedication; a man who, as Nancy knew, can be trusted. Time is short but Mun is willing to take some moments to reflect. He tells me that that morning's patient was a particularly tough one. She weighed more than four hundred pounds and had an inflamed gallbladder that needed to come out. Two other surgeons had turned her down for surgery because of her size, but Mun hadn't flinched. When he opened her up he found the largest gallstone he'd ever seen stuck to the wall of her gallbladder. He cut it out, and the woman lost three hundred cubic centimeters of blood. Most surgeons would have stopped there, but Mun went ahead with the bypass.

"I thought about putting it off, but I decided to go ahead," he said. "Gastric bypass was her only chance. This operation is not an elegant solution. On some levels it's barbaric. But, let's face it, it's all we've got."

Like Nancy, more than nine million American adults are "morbidly obese," roughly one hundred pounds or more overweight. Ten million more are almost there, teetering on the edge. But one need not reach this milestone to pass muster for surgery. People with "co-morbidities"—diabetes, heart disease, high blood pressure and the like—qualify at lower weights, and some physicians—among them, George Blackburn—argue that the bar should be lowered further, to include the still larger number of people who are obese, but not morbidly so. So many patients are clamoring for gastric bypass that experienced surgeons are overwhelmed. At a lecture I attended on surgical techniques at an annual meeting of the North American Association for the Study of Obesity, doctors overflowed the room, and stood straining to hear from the hall. The speakers, all obesity surgeons, promoted a slew of new techniques, including some that sounded only slightly less draconian than the old slice-and-dice approach. They also raised the specter of performing the operation on children as young as twelve. Such radical tactics are justified, they said, to stem the tide of an epidemic that is all but smothering us.

Nancy can't help but agree. She and I speak a month after her surgery, and she is cautious, but upbeat. She still cannot eat much, but then, she doesn't want much. She is living on nibbles of yogurt and scrambled eggs and instant breakfast mix, and hoping for the best. Gastric bypass patients lose most of the weight they lose quickly, in the first eighteen months, but that, Nancy knows, is the grace period. The hard part comes later, when the novelty of the procedure dims, and old habits start to surface. Often the pounds creep back. Somehow the magic wears off and the pouch gets stretched enough to accommodate the slowly sneaked calories. Or the person cheats by sipping high-calorie drinks all day long, or sucking down tubs of melted ice cream. Nancy knows that this could happen, and it worries her. But for now she is satisfied that the choice she made was the best one. For now, she is in control.

"I know this is no miracle cure," she says. "I know it's going to get rougher. I don't know what's going to happen two or three years down the road. This was a big step, and a terrifying one. But when it comes right down to it, I had no choice. It was my only chance at a life."

Two

WALKING NAKED

The Son of Man has come eating and drinking; and you say, "Behold, a glutton and a drunkard, a friend of tax collectors and sinners!"

—*Matthew 11:18, Luke 7:33*

. . . take no baths, sleep on a hard bed and walk naked as long as possible.

—Hippocrates, *On Diet*

Pudge is not a new thing. Consider the Venus of Willendorf, a hand-sized 25,000-year-old sculpture of oolite limestone. The Venus was unearthed early in the twentieth century in a cave excavation near the Austrian village whose name she bears. Her face is obscured by a decorative overlay of what looks very much like cornrows. She has erupting breasts, a Falstaffian stomach, and what appear to be no feet. She is lushly, voluptuously obese.

Willendorf is but the oldest in a long line of Paleolithic Venus figurines found scattered from southwestern France through Italy, Austria, and Turkey, to the north shore of the Black Sea. About one hundred such statues have been unearthed, and they are thought by archaeologists to be among humankind's earliest artworks. All these Venus statues, even the French ones, are grotesquely fat. Their marked similarities have prompted some anthropologists to conjure a continentwide "paleo-porn" cult, a delicious concept, but one not supported by evidence. Less fanciful observers consider them mere fertility goddesses, an interpretation that is probably closer to a truth we are likely never to know. What is certain is that artists of the period were realists, and that some Venuses show distinct knee abnormalities common in the super-obese. This makes it likely that the finely sculpted figures were modeled not from artistic inspiration, but from real life. Obesity, then, may have its roots in the Stone Age.

The Paleolithic era was a hardscrabble time. Glaciers spanned much of Europe, and the only reliable vegetation was moss, lichens, and grass, leaving humans to feed mostly on still-warm flesh ripped from the carcasses of other living creatures. Being alive meant being fit enough to scavenge or run down one's dinner. So the figurines present a puzzle: How did these women manage to grow fat in such lean times? Did adoring acolytes ply them with choice morsels? Did they trade favors for food? Whatever their tactics, the results were remarkable, for obesity was otherwise unheard-of in traditional hunter-gatherer societies.

It was not until the Neolithic revolution of ten thousand years ago that agriculture offered the first viable option to all that frantic stalking of prey and gnawing of flesh. Agriculture had the startling consequence of making food perennially available, first in the Fertile Crescent of the Middle East, then in China, Egypt, and western Europe. Agriculture also made possible the domestication of animals—and the steady consumption of animal fat. This bounty allowed a small elite to grow larger and fatter than they otherwise might have, though not all were eager to admit it. Egyptian pharaohs entombed themselves in chambers etched with depictions of their own gorgeous physiques, but studies of their mummified remains betray rolls of belly fat. (Servants were likely the living models for those hard-body renderings.)

The Greeks had some notorious gluttons, among them Dionysius, the tyrant of Heracleia who, in the era of Alexander the Great, terrorized much of what is now Italy. A seventeenth-century historian describes Dionysius as "an unusually fat man, which increased at length to such a degree that he could take no food which was not introduced into his stomach by artificial means." Whatever these mysterious "artificial means," they must have been marvelously effective, because eventually the monarch gorged himself into a state of such fatness that he could barely breathe. Plagued with sleep apnea, he frequently fell asleep on his throne in midpronouncement. "The Physicians ordered for remedy of this inconvenience, that Needles should be made very long and small, which when he fell into a sound sleep should be thrust through his sides into his belly. Which office his attendants performed, and till the Needle had passed quite through the fat, and came to the flesh itself, he lay like a stone; but when it came to the firm flesh, he felt it and awakened." Dionysius reportedly was not the least bothered by this, and in fact wished to die with his mouth full "rotting away in pleasure." Half a century later, Magas, king of Cyrene, a city-state and Greek colony near the North African coast, reportedly was granted a similar wish, being smothered in his own fat while lying in bed.

The ancient Greek elite had the resources and leisure to enable overeating, but by and large they had little tolerance for it. Hippocrates, known as the father of Western medicine (or at least one of the anonymous authors later known as Hippocrates) warned that "sudden death is more common in those who are naturally fat than in the lean." He advised "obese people with laxity of muscle and red complexion" to take their meals "after exertion and while still panting from fatigue and with no other refreshment before meals except wine, diluted and slightly cold. They should, moreover, eat only once a day and take no baths and sleep on a hard bed and walk naked as long as possible." Galen, the first-century Greek physician and anatomist whose ideas dominated Western medicine for a millennium, wrote that he could make a "sufficiently stout patient moderately thin in a short time by compelling him to do rapid running, then wiping off his perspiration with very soft or very rough muslin, and then massaging him maximally with diaphoretic injunctions, which the younger doctors customarily call restoratives." Ibn Sina, the tenth-century Arabic physician and tireless author of more than one hundred volumes of prose, lists obesity as a disease in his magnum opus *Kitab al-Qanun,* and suggests treating it with a regimen of hard exercise, lean food, and—contrary to Hippocrates' teaching—judiciously timed baths.

One need not read too closely between the lines attributed to these great scholars to discern a note of disdain in their injunctions. In Greece, birthplace of the Olympic games, allowing oneself to grow fat was tantamount to committing a crime against nature. In his still remarkably potent drama *Plutus,* the fifth-century B.C. satirist Aristophanes has the self-righteous and abstemious Poverty ranting, "But what you don't know is this, that men with me are worth more, both in mind and body, than with [wealthy] Plutus. With him they are gouty, big-bellied, heavy of limb and scandalously stout; with me they are thin, wasp-waisted, and terrible to the foe." The Cretans disdained the obese, and claimed to possess drugs (likely either toxic, purgative, or both) that allowed people to eat as much as they liked without growing fat. And despite their propensity for ritual gorging, the Romans, too, frowned on obesity. Roman women starved themselves, sometimes to death, in an effort to please their demanding husbands and fathers. The Spartans, never known for their tolerance, simply exiled their plumper citizens.

Given that Buddha is so often represented as a smiling obese man perched placidly in a lotus position, it is somewhat perplexing why, in parts of Asia, overeating has for centuries been viewed as a moral failing. A telling illustration of this is a twelfth-century Japanese *yamai-zoshi,* or picture-scroll of illness. In its original state, the scroll portrayed twenty-two disorders ranging from halitosis to hermaphroditism. It has since been cut into separate frames and one of these, an inked drawing of an obese woman supported by two hefty servant girls, hangs in the art museum of the city of Fukuoka. The accompanying text reads: "a woman moneylender who became exceedingly wealthy. Because she ate all kinds of rich foodstuffs, her body became fat and her flesh too abundant." The linking of moneylending with gluttony is not incidental—the woman is portrayed as greedy in every respect, and is paying for her heavy pocketbook with a bloated, ungainly body. That said, both the servant girls and a woman nursing a child in the background are overweight, so it may not be obesity per se, but the moneylender's gluttony, that the writer finds so objectionable.

Medieval Christian thinkers frowned on sensual pleasures, and it is no wonder that the list of seven deadly sins composed variously by St. Ambrose and St. Augustine ranked gluttony with lust, sloth, anger, envy, covetousness, and pride. In Christian teaching, gluttony is coupled with unworthiness—a condition, apparently, to which even the son of God was once (albeit erroneously) linked. In Matthew 11:19 and Luke 7:34, Jesus complains that unlike John the Baptist, who ate only honey and grasshoppers and never drank alcohol, he is sometimes mistaken by crowds as a *phagos*, or glutton, a "man given to eating," which is hard to believe given the emaciated figure he cuts in most renderings. The apostle St. Paul decried those "whose god is their belly" as having a sort of voluptuous moral deformity. In *Never Satisfied*, historian Hillel Schwartz points out that in medieval morality plays, "gluttony misled men to banquets in hell, where sauces were seasoned with sulfur and devils stuffed gluttons with toads from stinking rivers."

Obesity per se—the excess of fatty flesh—seemed to upset the European sensibility less than did the *act* of overeating. There are many possible explanations for this, but perhaps the most compelling is that the clergy were frequently far less trim than were their congregants—the fat priest appears so often in writings of the period (from Chaucer's hedonistic Monk to Robin Hood's companion, Friar Tuck) as to be a cliché. Poor sinners may have had the opportunity to gorge themselves after a hunt or a harvest, but they were far less likely to grow fat than were the well-fed clerics who preached to them. Not all gluttons were portrayed as fat in medieval imagery; indeed, some were downright scrawny. Preaching abstemiousness was clearly in the best interest of the clergy, who relied for their income on tithes from the common folk. For this reason it was perhaps prudent to disassociate the act of overeating with its physical manifestation.

The Renaissance and the Reformation brought a somewhat less narrow view of sin and a somewhat more scientific view of the workings of the human body. In the eighteenth century, some thirty doctoral theses (written, of course, in Latin) focused on the problem of obesity and its treatment, as did a number of monographs written in English. Most famous among these is *A Discourse on the Nature, Causes, and Cures of Corpulency*, by the Dutch physician, Malcolm Flemyng. Flemyng presented his treatise at a meeting of the Royal Society of Physicians in London in 1757. One can imagine that there was a good turnout for the talk, for in England concern over "corpulency" was on the rise.

Flemyng considered obesity a danger and an evil, but he did not dismiss the overweight as lazy, undisciplined, or sinful. Rather he portrayed them as unlucky inheritors of a predisposition that was not entirely within their power to control. Flemyng noted that some people seemed to put on weight more easily than did others, a quality to which he alluded vaguely as an inclination. He wrote: "Persons inclined to corpulency seldom think on reducing their size till they grow very bulky and then they scarce can or will use exercise enough to be remarkably serviceable."

Flemyng penned these lines more than a century before Darwin published *On the Origin of Species*, yet he reveals a grasp of a basic genetic principle—the idea that these "inclinations" were not twists of character, but could somehow be hard-wired into the human organism. The idea that particular characteristics are hereditary—that "like begets like"—is probably as old as humankind. But early notions of what constitutes the genetic material were sketchy. Aristotle proposed in the third century B.C. that every feature of every organism lay latent in the menstrual blood of the mother, and was activated by the sperm of the father. Roughly two millennia later, William Harvey, the British physician who became legendary for his charting of the human circulatory system, built on Aristotle's theory to suggest that all living things originate from the egg. Pierre Louis Moreau de Maupertuis, a French Newtonian scholar, proposed something of a compromise version of these theories in his book *Venus Physique,* published in 1745. Maupertuis anticipated many of the central ideas of Charles Darwin, suggesting, among other things, a rough theory of natural selection. He also proposed that males and females contributed equally to future generations. This concept was resisted by spermists, who clung to the idea that complete miniature organisms—"homunculi"—were poised waiting to be released into the mother from the sperm. (Imagine a tiny but fully formed human curled into a fetal position, awaiting the father's orgasmic sneeze.)

None of these early writers had any idea what it was that made horses give birth only to horses, or geese only to geese, or what caused siblings to resemble each other. But they did know that whether a man was tall or short or a woman was brunette or blonde had very little to do with the content of their characters and a great deal to do with their parents. Flemyng thought that corpulence, like height or hair color, might also have its genesis in something vaguely biological, and in this thinking he was, in a way, further advanced than many late-twentieth-century minds. He also had the good sense to recognize that morbid obesity was not the outward manifestation of a sinful inner core, but of an illness: "Corpulency, when in an extraordinary degree, may be reckoned a disease, as it in some measure obstructs the free exercise of the animal functions, and hath a tendency to shorten life, by paving the way to dangerous distempers."

Flemyng recounted the case of a forty-five-year-old physician, a small-boned gentleman of medium stature who, at nearly three hundred pounds, could walk no farther than a quarter mile at a stretch. Unable to complete his daily rounds on foot, the poor fellow was not only uncomfortable, but also in danger of losing his business. Not for a minute did Flemyng suggest that the doctor cut back on food or increase his pursuit of exercise, for he believed that the man's ingrained nature could not accommodate such changes: "[T]he habits of good eating and drinking become so deeply riveted as scarce to be conjured. A luxurious table, a keen appetite, and good company are temptations to exceed often too strong for human nature to resist."

Rather than attempt to change the very essence of the man, Flemyng prescribed a simple treatment: a daily draught of one-quarter ounce of castile soap dissolved in half a cup of water. The portly doctor lost nearly thirty pounds in "two or three" months on this regimen, and his health—and his business—Flemyng wrote, were enormously improved. Flemyng was delighted to pass along this happy news to colleagues, but, as a man of science, was hesitant to generalize from it: "I have taken opportunity of recommending the same remedy to others, in similar circumstances, but have not been as yet informed of any case in which a thorough trial thereof was made." Judging from the available literature of the period, Flemyng's cure met with less success than he had hoped, for obesity in Europe continued to pose a puzzle to the medical establishment and to produce a steady stream of bizarre case studies. Among these was Daniel Lambert, whose portrait adorns the foyer of the London Medical Society in leafy Cavendish Square. Lambert died in 1809, a few months short of his fortieth birthday. He weighed 739 pounds, and is said to have been an extremely happy man.

As public concern over obesity—or as it was then called, "polysarcia" (Greek for "much flesh")—continued to rise into the nineteenth century, so did the belief that this trend could be controlled, or even overturned, by assiduous effort. In his *Comments on Corpulency, Lineaments of Leanness, Mems on Diet and Dietetics,* published in 1829, William Wadd, surgeon extraordinary to the prince regent, attributed obesity to "an over-indulgence at the table," but unlike Flemyng, he did not believe that the behavior leading to this condition was immutable. (Perhaps it is dwelling on the obvious to wonder why Wadd, a hefty man, did not first heal himself.) Wadd's chief principle of treatment, "taking food that has little nutrition in it," drew on the work of the great German organic chemist Baron Justus von Liebig, a pioneer in physiological chemistry. Liebig's particular interest was agriculture; he argued convincingly that the growth of plants was dependent on certain soil nutrients, such as nitrogen, potassium, and phosphorus, and that the availability of these nutrients in the soil was the primary condition for plant growth. He also taught that body heat resulted from combustion of fats and carbohydrates, and was among the first to calculate the caloric values of food. From these calculations he concluded, "For the formation of body fat it is necessary that the materials be digested in greater quantity than is necessary to supply carbon to the respiration." Liebig speculated that only fat and carbohydrates were used by the body as fuel, and that protein was chiefly for the building and repair of lean tissue. Convinced of the primacy of protein, he went so far as to lend his name to two commercial products: Liebig's Infant Food, advertised as a replacement for breast milk, and Liebig's Fleisch Extract (meat extract), claiming that the consumption of these high-protein supplements would encourage the body to do "extra work."

Within a few years, French physiologist Claude Bernard showed that Liebig was only partially correct, and that protein could be both burned as fuel and converted into lipids to be stored as fat. The German physiologist Carl von Voit demonstrated the formation of lipids from protein in 1869, but the concept was so incredible as to be literally unbelievable to physicians of the time. To many the idea that protein could be transformed by the body into fat seemed no more plausible than gold being spun from flax, and von Voit's theory was disputed in lay and scientific circles for decades before coming into wide acceptance shortly after the turn of the last century.

No doubt the concept of the low-carbohydrate, high-protein diet so popular today had its grounding in Liebig's misperceptions, and in their early interpretations. Notable among these was a diet prescribed in 1862 by the British aural surgeon William Harvey for his friend and patient William Banting, an London undertaker whose fashionable clientele included the Duke of Wellington. Banting was slim as a young man, but in middle age grew so fat that he could not descend a staircase face first, for fear of being toppled by his copious stomach. Harvey had been to Paris and heard Claude Bernard lecture on the production of glucose by the liver, and the use of a "saccharine and farinaceous diet" to fatten up farm animals. He took from this that a diet high in carbohydrates was fattening and, conversely, that a high-protein, low-carbohydrate diet must be slimming. He prescribed the latter to Banting, who dropped thirty-five pounds in thirty-eight weeks on a diet of meat, small amounts of fruit, and liberal lashings of alcohol.

Banting's delight with this result moved him to compose and publish a monograph, *A Letter on Corpulence Addressed to the Public,* in which he wrote: "Of all the parasites that affect humanity I do not know of, nor can I imagine, any more distressing than that of Obesity." Banting enthused that his new eating and drinking regimen "cured" him of this "insidious creeping enemy," and recommended it to all who found bending over to tie their shoes a daily trial. Contributing to the diet plan's popularity was that on a typical day it made room for two to three glasses of good claret, sherry, or Madeira and a tumbler of grog, along with a "table spoonful of spirit" to soften up the dry rusks advised in place of the forbidden bread. Banting gave away 2,500 copies of his pamphlet to eager readers, and at the time of his death in 1878, nearly 60,000 more copies had been sold at sixpence apiece. By then, "Bantingism" had become a common synonym for dieting, and "bant" common usage for losing weight. Banting's *Letter* popularized the low-carbohydrate, high-protein regimen that is today ubiquitous.

The nineteenth century also brought the concept of "statistical medicine," the application of quantitative methods to the study of health and illness. Before then, medical students had been trained to use their senses to interpret the sight, sounds, and smell of disease. Clinical judgment relied on the elucidation of those senses, not necessarily on a systematic understanding of illness and its cure. But in the early 1800s physicians in France decided to put popular therapies to the test, and found most of them sorely lacking. To remedy this, the idea of studying disease in the population at large took hold. Vital statistics—the keeping of birth, death, and other records—became what historian Roy Porter called in his history of medicine, *The Greatest Benefit to Mankind,* a "thermometer of public health . . . involving the first systematic analyses of disease patterns." It suddenly dawned that health was best served not by the haphazard application of leeches, poultices, and potions, but by preventing disease in the first place.

A mastermind of this movement was Adolphe Quetelet, the Belgian dramatist, poet, astronomer, and mathematician now recognized as the father of social statistics. Quetelet cast doubt on the notion of free will, contending that humans were mere victims of statistical averages that, for example, preordained to a life of virtue or crime. This unsavory proposition paved the way for such unfortunate concepts as eugenics, but also cleared a path to more useful constructs, several integral to the broader understanding of disease. Quetelet is perhaps most famous for his concept of the *homme moyen,* or "average man." To reckon this average, Quetelet measured the chests of 5,738 Scottish soldiers and the heights of 100,000 French conscripts, and plotted them into a normal curve. From this he made the incidental discovery that at least 2,000 Frenchmen had lied about their height in an unsuccessful attempt to dodge the draft. Perhaps of more lasting interest was his observation that the weight of normal adults was proportional to their height squared.

Earlier in the century, physician John Forbes had written of the "impossibility of defining absolutely what degree of obesity is to be considered morbid." Quetelet's index solved this problem, offering a neat formula to delineate the merely stout from the truly fat. The Quetelet Index offered the first statistical measure of adiposity, one that a century or so later became the body mass index that is the gold standard today.

Meanwhile, the findings of Liebig and Bernard had caught the imagination of several other ambitious scientists. Among these was the young German physiologist Max Rubner, who in the 1880s built on their work to make quantitative determinations of the energy values of foods. Rubner put a respiratory apparatus into a doghouse-sized calorimeter (a device that measures energy expended as heat), stuck a mutt inside, and measured the heat production as a function of the dog's size and diet. His experiments determined that the exclusive source of heat in warm-blooded animals is the energy supplied by the food they eat. Rubner also discovered that metabolic rate within a species is proportional to the surface area of the body. For example, the metabolic rate of a 200-pound man is greater than that of a 150-pound man, simply because the larger man has more body area to support. Rubner also found that the metabolic rate increased immediately after eating, regardless of whether a person was chopping wood or just lolling around in bed nursing a box of chocolates. This "thermogenic" or heat-burning effect of food later became a serious preoccupation of obesity scientists.

Shortly after Rubner published his insights, Wilbur Olin Atwater, a chemist at Wesleyan University in Connecticut, published a series of articles on nutrition that popularized the concept of proper food constituents—fats, proteins, and carbohydrates—and offered scientific support for the idea that an improved diet might improve health. Atwater had a Ph.D. from Yale, where he had focused on the chemical composition of corn, and had studied with Rubner and von Voit in Germany and taught in Tennessee and Maine before settling into a professorship at Wesleyan. In 1894 he and his colleague E. B. Rosa, a physicist, built the first human calorimeter to directly measure metabolism.

In photos the calorimeter looks very much like a Lilliputian Airstream trailer—a four-by-seven-foot sealed room furnished with a sort of Murphy bed, a folding chair, and a stationary bicycle. The room was enclosed in double walls of sheet metal and entry made through a triple-glazed window. A test subject entered and either lay on the stiff little bed or pumped away at the stationary bike, while water was pumped through the chamber through circular tubing. The increase in water temperature before it entered and after it exited the chamber determined the amount of energy expended by the man inside. (J. C. Ware, a college athlete and bicycle racer, was Atwater's most famous test subject. He once expended 10,000 calories while cycling in place for sixteen hours.)

Using the German techniques he had mastered for measuring respiration and metabolism, Atwater computed the amount of fuel necessary to accomplish different levels of work. He estimated that men doing moderate physical labor needed 3,100 calories a day, and a whopping 120 to 130 grams of protein. He also calculated the energy value of food constituents—that a gram of protein or carbohydrate provides the human body with four calories, and a gram of fat with nine. He then drew up stern tables converting various foods into fuel units. The U.S. Department of Agriculture was highly intrigued by this concept, and in 1888 appointed Atwater the first director of the Office of Experiment Stations. From there, with the help of Andrew Carnegie's money and a bevy of "scientific cooks," Atwater managed to reset the nutritional standards for the nation.

Wesleyan was then a Methodist institution, and Atwater was encouraged by his superiors to apply his findings for the common good. The working classes of the time spent as much as 60 percent of their wages on food, leaving them little for education or housing. Having broken foods down into their basic components, Atwater knew that high-quality protein could be had in low-cost foods, such as beans, which he recommended as a substitute for meat. He did not expect his recommendations to affect the diets of the majority of Americans immediately. He wrote: "Of course the good wife and mother does not understand about protein and potential energy and the connection between the nutrition value of food and the price she pays for it, and doubtless she never will. But if knowledge is obtained and put in print, and diffused among those who have the time and training to get hold of it, the main facts will gradually work their way to the masses, who most need its benefit." The "masses" were not moved by this argument and protested, understandably, that calculating and prescribing the energy and protein needs of workers was demeaning. Labor leader Eugene V. Debs railed in *Locomotive Fireman's Magazine* that workers would resist "scientific degradation" and accused nutritionists such as Atwater of wishing to feed the U.S. laborer "at a cost as low as Chinamen are subjected to."

Atwater was indeed an elitist. He went well beyond the limits of his expertise to claim that workers became poor through their reckless "extravagance in food." Because he knew nothing of vitamins, he advocated that the masses forgo green vegetables and fruits, which he considered unnecessary indulgences, and fill up on cheap "wholesome" foods, such as wheat flour. Given that vitamins were not identified as such until 1912, perhaps Atwater can be forgiven such lapses. Among other things he was a social reformer, and like many of that breed, he believed that he knew what was best for others. He wrote, "If the . . . present waste of food material could be spent for more adequate shelter, the bad tenements in the slums would be renovated."

Atwater certainly lacked sociopolitical sensitivity, but aside from the gaffe over green vegetables, his science was largely sound. He made clear that food—whether it is protein, fat, starch, sugar, or alcohol—serves in the body to either form tissue or yield energy, or both. His work added much-needed scientific weight to a field rife with misunderstanding and quackery, and with Liebig he helped launch the country's first generation of nutrition scientists. Among these was Russell Chittenden, director of the Sheffield Scientific School at Yale University.

Chittenden was intense, driven, and frighteningly thin. In his text, *Nutrition of Man,* he waxed more evangelical than scientific, writing that "general over-feeding is a widespread evil, the marks of which are to be detected on all sides and in no uncertain fashion." He disagreed with von Voit and Atwater on the amount of protein necessary to keep the body healthy through hard work—he thought their estimates far too high. To test this hypothesis, he subjected a team of Yale gymnasts to a low-protein regimen.

After a year on what seemed a dangerously low protein intake, the athletes maintained their muscularity, and Chittenden had the photos to prove it. (It must be said that to these eyes the boys look awfully scrawny, particularly for "athletes," but Chittenden considered them splendid specimens.) Armed with this finding, Chittenden embarked on a campaign to let the world know the foolhardiness of overindulging in protein.

Chittenden seemed to have all the markings of an anorexic, insisting that slenderness per se was not proof that one was eating little enough. He writes: "While a superabundance of fat in the body is a sure telltale of overeating, the absence of obesity is by no means an indication that excess food is being avoided." Thin people do overeat, he insisted, but simply burn the excess calories more recklessly than do others, fueling a vast internal inferno that did them little good. To prevent this waste, Chittenden argued that food intake should not be regulated by anything as subjective and fleeting as hunger, but should be calculated scientifically by counting calories. Summing up his thoughts on this matter, he wrote: "Temperance in diet, like temperance in other matters, leads to good results. . . . [T]here is no demand on the part of the body for such quantities of food as customs and habit call for."

Thanks to the work of Chittenden, Atwater, and their colleagues, millions of Americans were introduced to the guilt-provoking notion that calories not only counted, but counted quite a lot. There was no more getting around the idea that the amount of food one ate determined, to at least some degree, the shape one assumed, and that the human metabolism operated under a set of inescapable natural laws. Rather than balk at this horrifying idea, the public enthusiastically embraced it. When Lulu Hunt Peters, the era's best-known female physician, penned *Diet and Health, with Key to the Calories* in 1917, the book sold an astonishing two million copies.

These advances did not prevent people from assuming, as many continue to assume, that there is something mysterious and magical in the accumulation of body fat. Well into the twentieth century, many nutrition scientists insisted that while calories certainly mattered, they did not matter enough to make all the difference between slim and obese. Like Chittenden, these experts assumed that people of similar body size and composition metabolized their food at radically different rates, that some were endowed from birth with slow metabolisms and thereby doomed to a life of either obesity or scanty portions, while others could gorge, fueling their raging internal flames with impunity. Atwater's discoveries somehow got lost in these assumptions, or at least put aside. His studies of energy metabolism were more focused on answering scientific questions than on public health, and were therefore perhaps easy to overlook. In any case, at the time, both the scientific community at large and the medical establishment had lost much of any interest they once had in obesity as a health problem. There were scattered articles on obesity in both European and American medical journals, but as cultural historian Peter Stearns points out in his comparison of French and American dieting traditions, *Fat History*, these did not add up to much. Physicians generally were not terribly vocal about the problem of overweight and, in fact, considered a modest padding to be beneficial, a safeguard against lean times. Obesity and overweight were of such little concern that doctors' offices were only rarely equipped with weighing scales.

Wise minds bicker over what precisely happened to, as Stearns writes, "put fat in the fire," but there was rather suddenly at the end of the nineteenth century a crescendo of critics denouncing fatness, railing against men who were "unduly stout" and women "overburdened with flesh." Some suggest that a growing interest in athleticism propelled the prevailing aesthetic toward a sleeker profile. Others say that professionalism rang the death knell for the "pleasingly plump"—in particular for women who flocked into the workforce in unprecedented numbers. But while these factors may explain the trend away from corsets, bustles, and other encumbrances of the Belle Epoque, they are unlikely to have prompted what was to become known as the "the century of svelte."

Truth be told, obesity (as opposed to Rubenesque plumpness) had not been fashionable since, perhaps, the Stone Age. French couturier Paul Poiret's "hobble skirt" and other clingy haute couture of the early twentieth century may have been more revealing than the fashions that preceded them, but not really much more than, for example, the waist-cinching styles of the Romantic era. While it is probable that the fashion industry, the proliferation of sport, and the entry of women into the workforce had something to do with the public outcry against fat, these were unlikely to be proximate forces. For it was not until the working classes could feel a strain at their belt buckles that a public campaign against fat began in earnest.

In the industrialized age, mechanization and mass production made life easier, in particular by making work less physically demanding. Food distribution and production became far more efficient, and food itself cheaper. Suddenly even common folk found a place at America's groaning board. Good Calvinists believed that prosperity was virtue's reward, but when so many could afford to eat well, portliness was no longer a badge of prosperity or, by extension, of virtue. The specter of fat as sinful was given all the more power when the rich no longer had a corner on it. And unlike drinking, gambling, or infidelity, obesity could not keep its own secret; it was a glaring manifestation of carnal appetites, a stick in the eye of the sanctimonious. As the working classes grew stout, any vestiges of sympathy for body fat were overwhelmed by an ardent campaign to root it out.

Underlying and reinforcing this movement was the new "science" that linked overweight with ill health and—the coup de grace—with a national financial burden. Brandeth Symonds, a physician employed by the insurance industry, wrote in 1909 that the relationship between body weight and health was now imbued with "commercial significance." He announced at a medical conference that all the old notions about the benefits of corpulence were bunk, and that husky types who claimed "it's all muscle" were fooling themselves. His colleague, Dr. Oscar H. Rogers of the New York Life Insurance Company, had already reported that men who were 30 percent or more overweight had an excess mortality of "34.5 percent," but Symonds took this one step further, warning that being as little as 10 percent overweight also shortened life. This was big news: men in the prime of life and at the height of their earning power were being cut down by overindulgence! This horrifying concept made headlines, and swayed even the lordly William Howard Taft to lose sixty-odd of his well over three hundred pounds before kicking off his presidential campaign (only to gain them all back and more during his first year in office).

Capitalism played no small part in the push toward thinness. Hillel Schwartz writes of "the double-edged threat of economic abundance: the exogenous fat man was the figure of overproduction, gluttony lured on by an economy run wild; the endogenous fat woman was the figure of underconsumption, domestic inefficiency in the midst of a floodtide of goods." Fatness was intolerable in an increasingly streamlined industrial age. It represented not only greed and excess, but, more importantly, inefficiency, sloth, and waste. It slowed people down and made them complacent, lethargic, and uninterested in getting ahead, when getting ahead was what being American was mostly about.

Hucksters hawked anti-fat nostrums like salvation—offering a little something not only to slim you down, but also to pep you up. According to a 1907 *New York Times* report, ads for pills, purgatives, and weight loss schemes were plastered "in newspapers and periodicals that take in such stuff, on the dead walls of old abandoned barns all over the country in letters three feet long [and] on the face of rocks visible from the passing railway in characters still larger, at all points where commendation of the remedy could be thrust into public view." Consumerism in this context was next to godliness, for no matter the cost, ridding oneself of unholy fat was like shaking off the devil. A space was cleared in every up-to-date middle-class bathroom for a home scale. Christine Terhune Herrick, in her book, *Lose Weight and Be Well,* published in 1917, called the scale a "materialized conscience" that weighed not only bodies, but worthiness—the lower the number, the loftier the soul. Weight tables were designed and redesigned, weight standards set and reset, national campaigns launched and relaunched in an attempt to hold back the inexorable wave of fat that American prosperity had made possible. As the century moved forward, calorie counting became a national—and international—obsession. Burning calories became a near-holy rite.

Thyroid extract derived from animals surfaced in the mid-1890s, to become if not the first, then certainly the most enduring medically prescribed obesity drug: physicians doled it out for the next seventy-five years. Thyroid medicine revved the metabolism, but it also triggered dangerously irregular heart rhythms, for which a chaser of arsenic, digitalis, or strychnine was often prescribed as a remedy. This was a dangerous business, of course, but no matter. One's duty was to rid oneself of loathsome fat; the rest was in God's hands.

Theories of the underactive thyroid's culpability in obesity persisted, perhaps because they were at once intuitive and vaguely "scientific." That fat people had a flickering metabolic flame fit neatly with the common wisdom linking fat to a lazy nature—turn up the flame, get those sluggish metabolisms moving, and solve the physical and moral problem. But into the twentieth century, as physicians gradually came to recognize that only a tiny fraction of obese patients suffered from a significant "thyroid inadequacy," attention began to shift from the butterfly-shaped gland in the neck to the large mass of gray matter in the skull.

Newly hatched notions of psychology brought to light the power of the unconscious to shape behavior, and of early life experience to distort and even pervert one's worldview. Freudian theory proposed that repressed drives—most of them sexual—could resurface in surprising places. Theorists on fatness took this to include the dinner table. The obese, it was said, had "fat personalities," a constellation of characteristics distinguished by an infantile need to suck and swallow. The obese ate instead of having sex, or, alternatively, ate and got fat to avoid having sex. The oral fixation theory led analysts to diagnose the obese as sexual cripples who sought release through what one German psychoanalyst dubbed "the alimentary orgasm." It was unclear whether this condition was particular to women, but perhaps because it defied credulity to suggest that men overeat to compensate for a sublimated libido, the language around this thinking had a distinctively feminine tone.

In the early 1950s, a few courageous members of the psychiatric community began to caution on the total dearth of hard evidence linking sexual maladjustment with fatness. Psychiatrist Harold Kaplan and his wife, clinical psychologist Helen Singer Kaplan, wrote: "Almost all conceivable psychological impulses and conflicts have been accused of causing overeating, and many symbolic meanings have been assigned to food." The Kaplans went on to list the more than two dozen rationales that analysts had served up as explanations for overeating—from satisfying a sadistic impulse, to penis envy, to pregnancy fantasies. None of these theories of "obesigenic" personality types were supported by actual evidence. The Kaplan review made devastatingly clear that obesity could not be blamed on any specific psychological pattern or malady.

Nonetheless, the psychological profession and the public continued to cling to the notion that the obese were anxious, unbalanced people who turned to food as a form of sublimation or escape. T. A. Rennie, an earlier thinker on the issue, put it most succinctly: "Obesity," he wrote, "can be regarded as a component of a neurosis, the physical expression of which is the accumulation of fat."

The assumed link between obesity and mental imbalance was understandable, considering that at the time the obese seen by psychotherapists often *were* neurotic. Though in hindsight it seems ridiculous that psychotherapists considered their patients a random sample of the population as a whole, this is roughly what occurred. At the same time, very few analysts were interested in obesity per se, considering the obese an undisciplined and undeserving lot, often stupid, and generally unworthy of the intense and glamorous attentions of psychoanalysis.

Albert Stunkard, a professor emeritus of psychiatry at the University of Pennsylvania, recalls those days vividly. Stunkard completed medical school and a residency in psychiatry at Johns Hopkins in the late 1940s. It was a time when psychoanalysis was not only the highest calling to which an ambitious psychotherapist could aspire, but pretty much the only one. Stunkard wanted in, and he selected as his mentor a man he now calls a "real Daddy," a decisive and imperious psychoanalyst of great stature and prestige. But after three years of living under Daddy's well-manicured thumb, Stunkard's respect for the discipline crumbled.

"Psychoanalysis was a very hierarchical field punctuated by the opinions of high-ranking people," he says. "I couldn't stand the elitism, or the ambiguity or the imprecision. There seemed to be no interest in data or in proof."

Stunkard left his training program at Johns Hopkins and headed north, eventually landing at Cornell–New York Hospital and the laboratory of neurologist Harold Wolff. Wolff was a serious scientist, the sort who had little patience for theories unsupported by empirical evidence. When Stunkard met Wolff, he was best known for his work on the psychological underpinning of disease—that is, on how emotion affects health. For example, through a series of well-documented experiments he had shown that prolonged emotional conflict could cause the stomach lining to engorge with blood and eventually to bleed. He was the first to prove that migraine headaches were not, as psychiatrists thought, "all in the head," but caused by dilation of the arteries of the brain. His overarching idea was that psychosomatic illness was the "inept" version of a normally "apt protective reaction pattern" that allowed humans to mobilize against stressful situations or events. "Psychoanalytic theorists (of the time) thought migraines were the manifestation of a hidden urge to kill your mother by destroying your head," Stunkard says. "It was no wonder I was so impressed by Wolff."

Stunkard talked his way into a fellowship in Wolff's lab, and, realizing he needed a project, decided almost arbitrarily to tackle a "small" problem suggested by an old medical school buddy, Theodore Van Itallie. At the time, Van Itallie worked at the Harvard School of Public Health with Jean Mayer, a highly regarded nutrition researcher who would one day become president of Tufts University. (Van Itallie later became a professor of medicine at Columbia University, and one the country's best known obesity specialists.) Mayer's research was fairly ecclectic, but by the mid-1950s, when Stunkard met him, he had focused his attention on obesity. He considered obesity a serious problem, but not a difficult one, and thought he had pretty much solved it with what he called the "glucostatic theory."

The glucostatic theory proposed that the obese had lost their ability to monitor levels of blood glucose, and that they were overeating due to the mistaken impression that the level was constantly low. Mayer believed that he had proved this theory in laboratory animals, and he suggested that Stunkard set aside three years to prove it true in humans.

"Mayer told me to spend one year studying food intake of mammals, one year studying disordered food intake in the obese, and take a third year to wrap it all up," Stunkard recalls. "Once we'd solved this obesity problem, he said that I could join him in a study of the physiological mechanisms of fever."

Stunkard never got to the fever part; he devoted the rest of his career in a noble effort to "wrap up" the obesity problem. Convinced after a year or two of research that low blood glucose levels were not a trigger for sustained overeating, he returned to the psychological literature, where he encountered the writings of Hilde Bruch. Bruch was a psychiatrist, but had spent her early career as a pediatrician with a special interest in obesity. She had begun practicing medicine in the 1930s, when glandular theories of obesity held sway. She found these theories lacking. "The error," she wrote, "lay in the application of mere endocrine labels which were simply repeated from one doctor to another, without being put to the objective test of validation." Nor was Bruch much impressed with the psychoanalytic focus on orality. Having seen many obese children and their families, she believed that the problem was rooted not in Freudian obsessions, but in a family dynamic in which the child was denied basic autonomy by parents who tried to control or pacify him with food. "I went to see Bruch in 1957, and she was very unhappy with psychoanalysis and hopeful that I would bring more empiricism to the field," Stunkard says. "I was very grateful to her for leading me in the right direction."

Applying Wolff's harsh empiricism to Bruch's questions, Stunkard was among the first psychiatrists to use standard scientific methods for studying the obese. He determined to sort out the specific characteristics of the so-called obese personality, and began by giving a battery of psychological tests to a group of eighteen obese men and another group of eighteen normal weight men. To his surprise, there seemed to be no consistent differences between these two groups. He attributed this to the sample's being too small, and turned to a data set collected years earlier for another purpose, a survey of 1600 residents of Manhattan's East Side. In this survey differences did emerge; fat people appeared slightly more neurotic than did thin people on three out of nine psychological measures: immaturity, suspiciousness, and rigidity. Still, the differences were slight, and not sustained in follow-up studies. Indeed, when Stunkard and others were finally able to amass enough data to make a scientific assessment of the question, they found no correlation between personality type and weight. Put simply, they found no evidence for the existence of an obese personality.

"In the early years, the respect for data in this field was absolutely lacking," Stunkard says. "Analysts made pronouncements and people believed them, that was all there was to it." Still, psychologists clung to the notion articulated so convincingly by Bruch, that the overweight tended to use food to self-medicate against depression or other psychological stresses. Bruch speculated that for fat people, food might be a potent antidepressant, using as evidence the fact that obese people had lower than average suicide rates.

(It apparently never occurred to Bruch that thin people may also turn to food for comfort in times of stress, but compensate for this later by eating less.)

Bruch's most enduring argument was that the obese could not recognize true physiological feelings of hunger and satiation—that they ate not in response to internal urges, but in order to fill some deep psychological gap. Stunkard tested this theory with a dramatic and vivid experiment in which obese and normal women swallowed balloons attached to a device that sensed stomach contractions. He found that unlike normal-weight women, obese women appeared not to recognize (or at least not admit to recognizing) stomach contractions as a hunger signal. This finding was never replicated, and it turned out to be misleading. But it nonetheless led to a series of experiments underpinning the widely held "externality theory" of obesity, a theory crafted by Columbia University psychologist Stanley Schachter.

Schachter was a charismatic, witty man, with a weakness for gossip, a bold sense of humor, and a voracious intellect. One of only a handful of social psychologists elected to the National Academy of Sciences, his interests ranged widely, but the binding theoretical theme of his career was the importance of social factors in shaping human perception of reality—for example, the power of environment to shape emotion. A graduate of Yale and the University of Michigan, he began his career at the University of Minnesota, and moved to Columbia in 1961, when he became interested in obesity. He performed a series of clever experiments, published as a series of articles in 1968, which he said demonstrated that the obese were far more likely than normal people to eat in response to environmental cues. His contention was that fat people were much more likely than normal-weight people to be prompted to eat by the sight of a luscious-looking chocolate cake, or by the time of day. Normal-weight people were less likely to respond to these cues, and more likely to wait for their stomachs to growl. This theory struck a chord with the public, and was widely reported in the press. The idea of fat people eating because they could not resist temptation rather than because they were hungry fit nicely with stereotypes of overweight people as weak-willed, hedonistic, and greedy. Like vanilla ice cream on a hot summer night, it was simply irresistible.

Schacter's externality theory was eventually proved wrong by one of his own students. Psychologist Judith Rodin (who in 1993 was appointed president of the University of Pennsylvania) reported in 1977 that obese people were no more likely than normal-weight people to eat in the absence of internal cues, or to eat more in the presence of external cues. Still, the idea had staying power. Hundreds of scholarly papers and popular books reinforced the idea that the overweight and obese were somehow different, and that curing obesity was a simple matter of reforming the behavior of obese individuals.

In the early 1980s, behavior modification became the gold standard of obesity therapy. Food was portrayed as dangerous, to be approached mindfully and with caution. Diet books proliferated, many of them using doctors' names in their titles to connote authority. Each had its own spin, but all took the position that constant vigilance was key. The overweight were encouraged to eat slowly and deliberately, and to keep records of their daily consumption. They were told to count calories, and then not to, told to reduce their consumption of fat or carbohydrates or, in some cases, even protein, and then told this was exactly wrong. And they were encouraged to approach their daily weigh-in with the trepidation of a prize fighter. In their *Psychology of Successful Weight Control,* Mary Catherine and Robert Tyson counseled that "you must regard the number on your scale as the most important number in your life."

Millions of people the world over took this edict to heart, some to frightening effect. In 1997, *Psychology Today* reported that 15 percent of women and 11 percent of men surveyed would sacrifice more than five years of their lives in exchange for achieving their desired weight. That's one in nine people willing to give up five years to die slim.

This seems a Faustian bargain, until one considers that the most viewed television show on the planet in 1997 was *Baywatch,* a flimsy beach drama in which gorgeous, pumped-up actors cavorted in micro-bikinis and spandex swim trucks. On *Baywatch,* the hard body was everything, and around the world, that idea had taken hold. Slimness was power, mastery, a ticket to success. The show made it seem easy. But of course it wasn't. We did not look like them. Here, on the other side of the looking glass, the world had never been fatter.

THREE

NATURAL BORN FREAKS

Some men by unalterable frame of their constitutions are stout.

—John Locke

The great blue whale, the largest mammal on earth, packs enough blubber on its carcass to stuff a typical bedroom from floor to ceiling. Yet fat accounts for only about 12 percent of its body weight. The fattest human on record, Jon Minnoch of Bainbridge Island, Washington, died in 1983 at age forty-two, dragging an estimated fourteen hundred pounds of flesh to his grave. Minnoch was about 80 percent fat.

The average middle-aged American lies somewhere between Minnoch and the whale.

Blue whales are the biggest mammals on earth, but, pound for pound, humans are by far the fattest. Which leads to the question: What, if anything, does all this fat do for us?

Fat is an organ, say scientists who study the stuff; think of it as you would, say, the kidney, the liver, or the heart. But it is difficult to think of fat in precisely those terms. For example, it is hard to imagine being on a waiting list for a fat transplant. Livers and kidneys are precious things doled out sparingly, one or two to a person. Fat seems to be in endless supply. Livers and kidneys are muscular, taut, useful. Fat is the color of jaundice, and has about the same appeal. But fat will fool you. For starters, it is not dispensable.

Under a microscope, fat cells look nothing like you would expect. Adipose tissue (from the Latin *adipatus,* meaning "greasy") appears orderly, even crisp, each cell a tiny hexagon cozied up to its neighbors in neat tessellation, like a honeycomb. Such clever distribution gives fat a spongy, bubble-wrap quality that makes it a great insulator and cushion. Fat holds in heat and absorbs shock. It protects joints and internal organs and eye sockets. And it is the ultimate energy storage system.

As human cells go, fat cells are enormous—thousands of times larger than most brain cells or red blood cells or the cells of the immune system that buffer the body from disease. Each cell contains a globule of oil, most of it triglyceride, a compound made up of fatty acid and glycerol. When food eaten chronically exceeds energy expended, most of the excess energy is stored as triglyceride in adipose tissue. One of the more amazing features of the mature fat cell is that it can grow or shrink in size severalfold depending on conditions.

Fat contains twice the energy-per-unit weight of protein or carbohydrate—nine kilocalories per gram of fat as compared with four kilocalories per gram of protein or carbohydrate. The average young woman carries on her body about one month's worth of energy as fat, which of course would be a very good thing were she, like her ancient ancestors, well out of supermarket range and foraging daily for roots and berries. It is a good thing, too, that fat is so energy-dense, because if this same young woman were reduced to storing in her body all that energy as carbohydrate, she would probably be too bulky to walk.

The mechanical properties of fat—as cushion, insulator, and fuel—have for some time seemed fairly impressive to those devoted to studying it. Doug Coleman is one of these. But Coleman differed from many scientists of his time in that he also believed that there was more than that to fat.

Douglas Coleman is in his seventies now, and retired, but he doesn't mind stirring up a bit of fuss, especially when the conversation turns to obesity science, to which, through no fault of his own, he happened to devote most of his career. He lives within a whiff of the Atlantic Ocean just outside Bar Harbor, Maine, in a beautiful home he built on a 196-acre plot of forest land. He spends a good chunk of time chopping wood there, as much to let off steam as to provide fuel for himself and for his neighbors.

Coleman was born and raised in Stratford, Ontario, about two hours' drive from Toronto. He was the first person in his family to attend college, working odd jobs to pay his way through McMaster University. "I worked for the railroad as a gandy dancer," he said—a laborer who lays down track. "But I did it so badly they gave me a different job." The job was not glamorous; he cleaned out half the urinals in Stratford.

Like his father, an appliance repairman, Coleman enjoyed taking things apart to figure out how they worked. The elder Coleman did this with radios and refrigerators, the younger with molecules. During graduate school in biochemistry at the University of Wisconsin, Coleman prided himself on being the fastest man on campus at identifying unknown organic compounds. Like all good scientists he was ambitious, but also cocky enough not to worry too much about a career path: he decided early on to follow the science wherever it led him. When he heard from a colleague that there was a job for a biochemist at Jackson Laboratories halfway across the country on Mount Desert Island in Bar Harbor, Maine he took it. He figured that it was as good a place as any to make a start, and that he would move on in a couple of years.

Bar Harbor is always a pleasure, and one can only imagine Coleman's delight when he first arrived. The village abuts resplendent Acadia National Park, where Rockefellers, Mellons, and Vanderbuilts once kept Gatsby-style "cottages" and trolled in yachts the size of aircraft carriers. Though many swells have since moved on to parts more exclusive, Bar

Harbor remains a blue-chip vacation destination. Coleman could never bring himself to leave the place, even when things turned sour.

Coleman had come to Jackson to study the new field of biochemical genetics, specifically how genetic information gets converted into proteins. Proteins are built on a gene template, and genes are built of DNA. It seems ages since the DNA structure, the double helix, was first elucidated by James Watson and Francis Crick. But Coleman can remember the event easily. It happened in 1953, just a year after he earned his bachelor of science degree, and a hundred or so years after an obscure Moravian monk, Gregor Johann Mendel, first published the results of his seminal pea experiments, which set in stone the rules of genetic inheritance by demonstrating that pairs of "factors" (later termed genes) determined the inheritance of traits. Some traits, Mendel found, were dominant, meaning that they showed up in every succeeding generation. Others were recessive, and could be passed from generation to generation without notable effect, manifesting only when transmitted in a double load, one gene from each parent.

Mendel's findings were stunning, but so far ahead of their time that scientists didn't pay them much mind. Like a wheat field in winter, the field of genetics essentially lay dormant until 1900, when Mendel's work was rediscovered almost simultaneously by three botanists, and made public by the British biologist and Mendel evangelist William Bateson. A few years later, British physician Archibald Garrod proposed that each gene bore with it the chemical recipe for a single protein. He came to this simple but astonishingly profound conclusion as a clinician, by observing the course of alkaptonuria (AKU), a rare and usually mild disorder that darkens the urine and often leads to arthritis. AKU occurs when the body is unable to rid itself of a substance called homogentisic acid. Garrod had observed that AKU passed from one generation to the next, just as predicted by Mendel's rules of inheritance for recessive traits. From this he deduced that something must be missing in patients who inherited double copies of the recessive gene, a glitch he famously described as an "inborn error of metabolism." In the 1908 Croonian Lectures to the Royal College of Physicians, he proposed that AKU was caused by a missing enzyme, a protein catalyst that normally helps the body break down and get rid of homogentisic acid. He went on to speculate that other disorders, like albinism, might share the same mechanism, and proposed a relationship between genes and proteins that suggested chemical individuality as the paradigm of Mendelian variation.

Garrod's thinking was prescient but, like Mendel's, too far ahead of its time to have an immediate impact. It would be well into the twentieth century—1941—before American scientists George W. Beadle and Edward Tatum refined Garrod's thinking into the "one gene, one enzyme" hypothesis that explained how genetic information is converted into proteins. Three years later, in 1944, scientists finally agreed—after years of acrimonious debate—that the genetic material itself was DNA, a curiously simple molecule composed of sugar, phosphate, and four nucleic acid bases: adenine, cytosine, guanine, and thymine, A, C, G, and T. And just short of a decade after that came the discovery of DNA's equally simple architecture—the iconic double helix. From this discovery Watson and Crick were able to explain how the molecule transmitted instructions from one generation to the next. As Crick and others deduced, the sequence of the DNA bases—A, C, G, T—spell out the instructions for the synthesis of proteins. In humans, three billion of these letters, coiled like spaghetti, assemble into our roughly 30,000 to 40,000 genes. The letters are assembled linearly along DNA strands on twenty-three chromosome pairs. Together, these fragile-looking strands make up the sum of our individual potential.

The human genome can be compared to an encyclopedia, in which the twenty-three pairs of chromosomes are volumes. Each volume contains several thousand entries or genes, and each entry has many paragraphs containing thousands of words written in three-letter "codons." In the early 1970s Doug Coleman was one of hundreds of scientists straining to make sense of these entries. What Coleman wanted to know was what most molecular biologists wanted to know: which genes—which particular configurations of As, Cs, Gs, and Ts—coded for which traits.

The totality of our genetic code—the genome—is often described as a blueprint, but that's not quite it. The genome is really a set of opportunities. All the information needed to build a cell is contained in the genes, but as scholars have pointed out, this simplistic statement begs the question of what is meant by "information." This is not a philosophical quibble, but a serious and practical question. Because genes are turned on and off by the circumstances of the cell around them, a "genetic program" is not set in stone. Like light bulbs, individual genes must be turned on or "expressed" to have an impact—they must be translated into proteins. Genes are not always expressed. The likelihood that a gene will be expressed physically in the resulting organism is called penetrance, and penetrance will vary from gene to gene and from organism to organism. Genes that are not expressed are not material to our development. This explains why identical twins, who share the same genetic material, are never exactly alike—not all the genes they inherited are expressed in precisely the same way. It also explains why people may have a particular gene, yet not display the physical characteristic associated with that gene. This is why, for example, at least 20 percent of women who carry even the most menacing breast cancer gene never contract the disease.

The genetic material is fundamental. It provides something for the environment to act upon. Irregularities in the gene—a switch of a G for an A, for example, or missing letters or additional letters—can have an enormous impact, just as the meaning of a word or sentence or phrase can be changed, or even reversed, by a slip of the tongue or pen. Consider the innocuous, though perhaps annoying, sentence: "He enjoyed a nice pair of buns as much as the next person." Now trade the "b" in buns for a "g," and the sentence remains annoying, but in a completely different way. An equally small typo in the genetic code can have equally profound consequences.

* * *

Coleman's early focus was not on obesity, but on looking for typos in the genetic code that result in muscular dystrophy, a devastating muscle-wasting disease. The search began, as do most quests at Jackson Laboratories, in mice. Jackson was and is a sort of mouse ranch, squirming with rodents both ordinary and fantastic. There are today 2,500 different strains breeding in forty-seven "mouse rooms" scattered around the Bar Harbor campus.

I came to Jackson to speak with Coleman in late fall, when the leaves had turned and the tourists had returned to their homes in Boston, Los Angeles, and London. The laboratory sits high on a bluff overlooking the Atlantic Ocean. It is low-slung and built of red brick, a bit off the tourist path. This is probably a good thing, for aside from its views, it is not a charming place. The mouse house was off limits to visitors, tucked behind the main buildings with windows shrouded; I was told this was a precaution taken to ward off animal rights activists. I did manage a glimpse through a half-open shade when my minder's back was turned. (Like the mice, visitors here are monitored by authority figures, albeit kind and helpful ones.) What I saw was a room stacked floor to ceiling with what looked like Tupperware containers, about as innocuous a sight as one could imagine. Of course, one could also imagine the mice inside that Tupperware, running in tiny circles and awaiting their fate.

Jackson is a living library, and rodents here are most valued for their potential to yield information. The weirder the mouse, the better. Consider the "Rhino," a hairless albino mouse shriveled as a raisin. Shivering and sniveling, the rhino wouldn't last a day in the natural world, but immunologists pay eighty-five dollars for a breeding pair. Another strain, "Flaky," has thick, scaly skin, and is used in studies of psoriasis. "Stargazers" have an inborn defect of the inner ear that condemns them to crane their necks ever skyward, like a bird-watcher possessed. "Punk Rockers" are headbangers, and "Waltzers" dash around their cages in an endless figure-eight. In photos, Stargazers, Punk Rockers, and Waltzers appear to have the manic quality of psychotics off their meds. Flaky and Rhino look merely pathetic. Next to this assortment of sideshow freaks, the plump and placid *obese* mouse looks regal.

Mouse mutants have for centuries been an object of fascination. The ancient Cretans are reputed to have kept a colony of ghostly white, pink-eyed albinos in a temple dedicated to the mouse god, Apollo Smintheus. Chinese emperors included a selection of mutant mice in their well-stocked menageries. And in the nineteenth century, Asian breeders developed a variety of mutant mouse lines and shipped them off to "mouse fanciers" in Europe and America. Among these was Abbie Lathrop, a retired school-teacher who bred mice for sale as pets from her home in Granby, Massachusetts. Lathrop lived near the Bussey Institute, a scientific outpost of Harvard University directed by professor of biology William Ernest Castle. Castle bought some of Lathrop's fancy mice in 1902, and with his many students began a systematic analysis of their inheritance and genetic variation.

One of Castle's students, Clarence Cook Little, developed the first recognized inbred mouse strain, which was used to create the genetically uniform "A strain" distinguished by its propensity for growing tumors. In 1929, Little and seven of his colleagues brought this and other mutants to Bar Harbor. The new laboratory was named for one of its chief benefactors, Bar Harbor summer resident Roscoe B. Jackson, the recently deceased head of the Hudson Motorcar Company. Cancer research was the laboratory's focus, and the mouse its experimental model. In an interview with the *Bar Harbor Times,* Little explained that "there will be 50 or more strains, or families, of mice, each differing from others in the degree to which cancer appears and the type of cancer which arises." It is unclear whether the local citizenry cheered this prospect as they hurried off to check their lobster traps.

Mice do not in any way resemble humans, but they are similar enough genetically to offer a reasonable model for the study of normal and abnormal human biology, as well as human disease. Little believed this, but at the time, such thinking bordered on heresy. Only four years earlier, William Jennings Bryan, the three-time Democratic candidate for president, had led a fundamentalist campaign to ban the teaching of evolution in American classrooms. To suggest that humans had anything in common with chimps or apes, let alone mice, was to risk a fistfight in some quarters. But Little persisted, the laboratory was built, and the inbred mouse proved a remarkable tool for research in hereditary disease. By the late 1930s Jackson was not only breeding stock for its own use, but also making a substantial profit by selling mice to other laboratories.

In 1947 a wildfire swept through Mount Desert Island and the laboratory, incinerating all but a scattering of the mice. Little was determined to rebuild, and donations of mice—all of them originally bred at Jackson—poured back to the lab from around the United States, Canada, and Great Britain. Among these was a new mutant, the dystrophic mouse that Coleman would use as his model for the study of muscular dystrophy. And two years later, another mutant suddenly appeared in the lab—a mouse with traits that would, some twenty years later, attract and hold Coleman's attention for the rest of his career.

An animal caretaker first spotted the creature huddled in a corner of its cage, grooming itself. It was furrier than most, but what really stood out was the size of the thing—it was hugely fat. The caretaker alerted doctoral candidate Margaret Dickie, who diagnosed the mouse as "pregnant." But there were problems with this theory. For one thing, the mouse never delivered a baby. And on closer inspection, it turned out to be male. The fat mouse ate three times the chow eaten by a normal mouse, pawing for hours at the bar of the food dispenser like an embittered gambler banging away at a recalcitrant slot machine. Between feedings it sat inert. It seemed to have been placed on this earth for no other purpose than to grow fat.

There had been other fat mice. The agouti mouse, named for its mottled yellow fur similar to that of the burrowing South American rodent, is, in its "lethal yellow" mutation, double the weight of the ordinary variety. But the fat agouti was svelte compared to the newcomer. This mouse was outlandish, a joke, a blob of fur splayed out on four dainty paws like a blimp on tricycle wheels. Rather than dart around the cage in mousy abandon, it was docile, phlegmatic, as though resigned to some unspeakable fate. Dickie and her colleagues christened the new mouse "*obese,*" later abbreviated to "*ob,*" and pronounced "O.B.," each letter drawn out in its own languid syllable.

As Coleman tells the story, a number of investigators at the time took a stab at studying the *ob* mouse. They were foiled on several counts. The *ob* was extremely difficult to breed—for reasons unknown at the time, the females were infertile, so their ova had to be painstakingly transplanted into normal females. Complicating things further, the effect of the *ob* mutation varied depending on what particular strain of mouse carried it. As a result, researchers working with one strain of *ob* mice got findings that were often contradicted by researchers working with another strain. This was terribly frustrating, and eventually the *ob* mouse fell out of favor as a study subject.

Shortly after Coleman began working at Jackson, yet another fat mouse cropped up. This mouse was as fat as the agouti yellow mutant, but thinner than *ob*—a linebacker, not a sumo wrestler. What distinguished it from the other mice was its drinking habits: it drank about twenty times the water consumed by a normal mouse, and urinated so much that it stank up the lab. This caught the nose and the attention of Jackson scientist Katharine Hummel, a stern and demanding Minnesota native who specialized in diabetes research. Hummel had a hunch that the mice were suffering from diabetes, and she asked Coleman to help her to look into it. Coleman had no idea what he was getting into, but being at that point a sort of scientist-of-all-trades, he agreed.

The new mutant turned out to have very high blood and urine glucose levels due to a pancreas that was unable to properly produce insulin. Officially diabetic, Hummel christened it as such, but had the delicacy to also give it a nickname, *db.* Because the *db* mouse resembled the *ob* mouse in being fat and sterile, Hummel and Coleman assumed that the *db* mutation would be found on the same gene as the *ob* mutation. It would be twenty years before scientists could actually locate a gene on a chromosome and make copies or "clones" in bacteria. But even in the late 1960s, the dark ages of molecular biology, techniques existed to determine whether two mutations shared the same gene.

Both *db* and *ob* are recessive mutations, which means that a mouse must inherit the defective gene from each of its parents to acquire the condition. Coleman crossbred the carriers of *db* with carriers of *ob*, to determine which, if any, of their offspring manifested signs of either mutation. If *ob* and *db* shared the same gene, then by the laws of genetics one in four of their offspring would be fat. None were. The *db* and *ob* defects were carried on different genes, and, as it later came clear, on different chromosomes. The *db* was an entirely new mutation.

Coleman became fascinated by these mutants, and obsessed with what they could tell him about human disease. Other scientists, too, were deeply intrigued. Diabetes and obesity were widespread, and any clues to their etiology were eagerly awaited.

"I gave a talk in Kalamazoo, Michigan, and people stayed and asked questions for two hours," Coleman recalls. "People told me that our mutants would revolutionize diabetes research. I was invited to Paris—Paris! This was in 1968, and to be in Paris, during the student riots, that was something, a really big deal for me."

Though by nature a modest man, Coleman couldn't help but be flattered. Suddenly he was the toast of the international lecture circuit, a personage among scientists. These mutants, it seemed, would be his ticket, and he focused his sights on learning everything he could about them. He found that *db* possessed a number of puzzling traits. For one thing, it could go days without food. For another, it was resistant to the blood sugar–lowering hormone insulin, so much so that it barely flinched from an injection large enough to kill an ordinary mouse. The *db* also had unnaturally high blood insulin levels, and Coleman speculated that some unknown "insulin-releasing factor" was circulating in its blood. Searching the literature, he found a method to test this theory. It was called parabiosis.

Parabiosis is by any standard unsettling, and by animal rights standards sadistic. It involves slicing the skin of a mouse two-thirds of the way down its body—a two-inch slash on a three-inch mouse—and piercing its peritoneum, the membrane lining the abdominal cavity. The same slash and piercing are made in a second mouse, and the two are sewn together, like Siamese twins. The conjoined mice stumble off in a daze, and somehow manage to keep their acts together while the wound heals. Within a few days, the skin and peritoneum of the mice grow together, and their circulatory systems link through the sprouting of small blood vessels. Every minute, about 1 percent of their blood is exchanged through the vessels, in a perennial transfusion that allows two mice to essentially share one blood supply. The purpose of this procedure is to learn what, if anything, in the bloodstream of one animal will affect the health or behavior of the other.

Coleman paired a *db* mouse with a normal mouse, to see whether the *db* mouse carried in its blood some substance that caused its insulin to spike. Were this the case, the insulin-triggering factor in the fat *db* mouse would pass through to the normal mouse, and elevate the normal mouse's blood insulin levels. Conversely, were there an insulin-suppressing factor in the normal mouse, this would pass into the diabetic mouse and lower its blood insulin levels. Either one of these outcomes would have been a spectacular scientific find, as it would indicate that some as-yet-unknown factor, probably a hormone, underlay diabetes in these mutants. But neither occurred. Instead, something far less exciting happened: the mice died.

Coleman considered himself to be a rather clumsy surgeon, and his first thought was that he'd botched the operation. So he did it again, and then a third and fourth time. The mice died. This was puzzling. After a rocky start, Coleman had grown fairly confident in his surgical technique. Examining the dead pairs closely, he saw no sign of surgical error, or that the two strains were rejecting each other immunologically. But he did notice that the normal mice had no body fat, and no food in their stomachs or intestines. He began closely watching newly joined pairs, and observed that the normal mice never seemed to eat. Clearly, the normal mice were starving to death, and, because they shared blood systems, were bringing the *db* mice down with them. The question was, why? Coleman turned to the scientific literature for clues.

Starting in the nineteenth century, scientists studied the brains of laboratory animals by destroying tiny bits and noting the impact. For instance, cutting or burning a lesion in the cerebellum of a dog made it difficult for the animal to walk, offering evidence of that brain region's importance in voluntary muscular activity. A cat with a lesion in its parietal lobe might lose part or all of its vision, indicating that this part of the brain was somehow involved in sight. In 1943, the American neuropsychologist John Raymond Brobeck discovered that lesions in the hypothalamus, a pea-sized region in the base of the brain, caused hyperphagia or uncontrolled eating in rats. Brobeck didn't have a clue as to why, but a dozen years later British scientist G. R. Hervey worked out a partial answer. Hervey lesioned the hypothalamus of rats, and then conjoined them to unharmed rats. The lesioned rats ate a lot and became fat, while their partners ate very little and grew thin. When Hervey made lesions in the hypothalamus of the thin rats, their appetites surged as well. "It is therefore suggested," Hervey wrote, "that the partners of animals with lesions most probably became thin because their own hypothalamic controlling centres reduced their intakes of food. . . . This implies that some influence or 'information' from an overfed body can affect an intact hypothalamus in such a way that feeding is reduced."

Another British scientist, Gordon C. Kennedy, had done a series of elegant experiments a few years earlier that pointed to something in the fat cells themselves that directly or indirectly controlled feeding in rodents. He concluded from this that body weight in rodents is controlled by a sort of fat thermostat—which he christened a "lipostat"—that senses how much fat there is on the body and adjusts eating and energy expenditure accordingly to maintain a steady state or "set-point." Hervey built on Kennedy's thinking to suggest that the hypothalamic region might somehow intervene in this feedback mechanism by picking up a satiety signal conveyed from the fat through the bloodstream. The lesioned rats were immune to the signal, while the normal rats sensed it all too clearly, for they stopped eating and died.

Noting the thinking of these scientists, Coleman set out to repeat his experiment, this time in both *ob* and *db* mice. When he grafted the *ob* mouse to the *db* mouse, both mice grew thin and died, just as the normal mice had done in his study three years earlier. When he paired an *ob* mouse with a normal mouse, both mice did fine, though the *ob* mouse did not get as fat as expected. And when he paired an *ob* mouse with another *ob* mouse, both mice stayed fat.

Coleman published these findings in 1973 in *Diabetologia*, a journal read chiefly by endocrinologists. The article described the experiments and offered a tantalizing explanation. Coleman theorized that normal mice produce a satiety factor that travels through the blood to the brain, and signals the mouse to stop eating. Obese mice, he theorized, lack this factor. Diabetic mice make the factor, but lack the equipment—the receptor—to sense it. Hence, some unknown molecule circulating in the blood and mediated by the hypothalamus was signaling rodents when and when not to eat.

Like Hervey and his followers, Coleman believed that eating behavior in rodents was controlled directly by a factor in their blood. But he elaborated on this to suggest that eating behavior might be regulated to some degree by a biological feedback loop controlled by one or more genes. All this, of course, made perfect sense. It was supported by reproducible evidence, and grounded in historical precedent. It also offered the first compelling biological evidence of what so many scientists—and lay people—had observed: that body weight seemed to be "set" by some mechanism that eluded willful control. In short, it was a breakthrough, and by far the most important finding to date in obesity science. But, like Mendel's discovery of the genetic material, it was too far ahead of its time to be taken seriously by other scientists.

The American philosopher of science Thomas Kuhn argued in *The Structure of Scientific Revolutions* that scientific "knowledge" is not absolute but relative, dependent on the paradigm or theoretical framework that dominates the field. Such paradigms are so dominant, Kuhn argued, that regardless of countervailing evidence they are accepted as gospel until a "scientific revolution" creates a new orthodoxy. Farmers and rangers had long known that farm animals could be selectively bred to eat more and grow fat, but the idea that human eating behavior could be influenced by biology, and that this biology could be directed by genes, was at the time not palatable. Humans, after all, were not livestock.

When Coleman published his findings in the early 1970s, the prevailing scientific view claimed obesity as a problem of behavior, not biology. Behaviorists who had cut their teeth on experiments with pigeons and dogs had flocked in droves to treat overweight humans with behavior modification, in part because the results—pounds lost—were so concrete, so tangible, so easy to measure, even if those pounds were quickly regained. Behavioral approaches also had a galvanizing entrepreneurial allure—they were cheap to run and attracted plenty of customers. Weight "treatment centers" appeared around the country, claiming to "deprogram" overeating by the judicious application of everything from rewards to electric shocks.

Claims aside, behavioral approaches were not working for most people. The underlying—though rarely voiced—problem was that behaviorists couldn't get down to the bottom of things: they couldn't explain why some people seemed less inclined to eat and to get fat than did others. Chalking it all up to "bad" or "unhealthy" behavior begged the question of just what underlay the behavior, and why some people and some animals seemed to develop these "bad habits" in a given environment when others did not.

As Albert Stunkard wrote in a 1974 address: "[B]ehavior modification may simply be helping someone who biologically should be obese to live in a semistarved state." In 1979, Stunkard and a coauthor, Sydnor Penic, reported on a five-year follow-up of patients treated in the first major behavior modification program. The results, they politely concluded, were "disappointing."

Coleman's findings offered a tantalizing explanation of why behavioral approaches were failing, but few people considered them seriously. He stuck staunchly to his theory nonetheless, and continued to search for the mysterious satiety factor for the rest of his career. One man, a physician whom Coleman declines to name, claimed to have purified a factor from the pancreas that he injected into a *db* mouse, curing it of its diabetes and obesity. For years, Coleman tried to duplicate this experiment, to no avail. Finally, Coleman and his colleagues invited the physician to deliver a paper detailing his results. The doctor flew to the meeting in his own private plane. He hung around for a while, and then suddenly disappeared.

"The guy was a liar and a cheat; he had falsified his papers," Coleman says. "And he set back the whole field."

Meanwhile, Coleman's own best instincts were leading him astray. For years he thought fatty acids—which, with glycerol, make up the blood triglycerides—carried the satiety message. Given that fatty acids are made in fat cells and circulated in the blood, this was a logical assumption. But it was wrong.

Doug Coleman retired from Jackson Laboratories in 1991, and he made a clean break. He'd spent most of his life in science, and he had seen and heard enough. At his farewell party, someone recited the poem "Two Tramps in Mud Time," by Robert Frost. Coleman recalls the first stanza in particular:

> Out of the mud two strangers came
> And caught me splitting wood in the yard.
> And one of them put me off my aim

By hailing cheerily "Hit them hard!"
I knew pretty well why he dropped behind
And let the other go on a way.
I knew pretty well what he had in mind:
He wanted to take my job for pay.

Coleman in some ways resembles Frost, not so much in looks, but in his actions, which he much prefers to words. He swaps wood from his lot for venison, or gives it to neighbors who help him haul brush. He left Jackson not to retire, but to move on to a new mission, to forest management and land protection in his adopted state of Maine. He feels a deep bond with the land. Like Frost, he has grown crusty, and just a touch bitter.

Coleman is not the sort of man to have regrets, but if he were he would regret this: not having the genetic tools to ferret out that elusive satiety factor. Three years after his retirement party it was found, by a man who called Coleman his friend. This man was not the sort to barter firewood for deer meat, or to spend long hours hacking through underbrush to clear recreational trails for strangers to wander. This was a very different sort of man; a very different sort of scientist—the new sort.

Four

ON THE CUTTING EDGE

> Nothing in the world can take the place of persistence. Talent will not; nothing is more common than unsuccessful men of talent. Genius will not; unrewarded genius is almost a proverb. Education will not; the world is full of educated derelicts. Persistence and determination alone are omnipotent.
>
> —Calvin Coolidge, 1932

On a damp morning in April, Jeffrey Friedman's laboratory at Rockefeller University is dreary to the verge of depressing. There is a vague odor of mouse droppings and the dull, almost subthreshold drone of laboratory instruments measuring things. A huddle of rubber tree plants on a nearby window sill offers forlorn evidence that the workers here—postdoctoral fellows, graduate students, laboratory technicians—lack either the time or the will to nurture. An unmanned computer monitor flashes a screen saver of someone's chubby newborn, but otherwise there is little that is personal here, little to distract from the business at hand. It took Friedman eight years of round-the-clock vigilance to track down Coleman's elusive satiety gene, and he demands no less from his acolytes.

Jeff Friedman holds a chaired professorship at Rockefeller, where he is head of the Laboratory of Molecular Genetics, director of the Starr Center for Human Genetics, and investigator at the Howard Hughes Medical Institute. A tall man, with a tall man's gently sloping shoulders, he has a tangle of dark hair, a mustache, and hands big enough to palm a basketball. He plays ice hockey on the odd Sunday afternoon, and tennis when he can, but his body looks underused, like the road bike gathering dust in the corner of his office. A colleague once described him as "nerdy and gauche," but if that were ever the case, it is no longer. In middle age he has become generous, even gracious, and has cultivated an interest in art and the opera. He is an eloquent public speaker, and tells a decent joke. But he has yet to outgrow his sometimes off-putting impulse to be the smartest person in the room. He is not always interested in the thoughts of others.

That Friedman was precocious should go without saying; at Rockefeller precocity is prerequisite, the price of admission. What sets him apart is determination. He is tireless, dogged, a master of persistence, a man who can delay gratification almost indefinitely. One friend put it this way: "Jeff has an overabundance of willpower." Others see it differently. They say Friedman is to ambition what ocean is to wet.

Friedman was born in Orlando, Florida, but as a kid he never made it to Disney World. He moved before it was built, to Woodmere, Long Island, a mostly prosperous bedroom community favored by families from Brooklyn and Queens moving on to the suburbs. Like his father, a doctor, and his brother, a doctor, Jeff was expected to achieve. Unlike them, he did so strictly on his own terms. "What stands out is that when Jeff was told he couldn't do something, or that he wasn't allowed to do something, he did it," his brother Scott told me. Friedman was not a stellar high school student. Among his few enduring memories of that time is writing a paper on the plight of Native Americans for which he appropriated most of the information from the left-wing magazine *Ramparts.* His high school biology teacher has a vague recollection of a "tall, thin guy, with curly hair, not a standout academically, a bit of a joker." There is nothing at all written about him under his high school yearbook photo—no honors, no clubs, no team affiliations.

Friedman polished off high school at age sixteen, undergraduate and medical school at age twenty-two, and a residency in internal medicine at

Albany Medical Center at age twenty-five. He was appointed chief resident, which is considered a serious honor. But he was bored, restless. Clinical medicine wasn't working for him—in part, former Rockefeller colleagues say privately, because he didn't have the patience for patients. A lightning-quick intelligence paved his way through medical school, but it didn't prepare him for the tediousness of medical practice, or the messiness of dealing with an aggrieved parent whose child had died. Clinical medicine didn't deliver much in the way of intellectual thrills. It was basically the same thing day after day: listening to patients, probing patients, treating patients, and, worst of all, standing by helplessly when there was nothing more to be done. He hated that—the standing around, the powerlessness, the hopelessness. When patients improved there was gratitude, but gratitude was not what he wanted. He longed for challenge and control and he believed he would find these things in the increasingly glamorous field of molecular biology. When a planned clinical fellowship was postponed and a medical school mentor offered to arrange a research fellowship at Rockefeller University, one of the preeminent medical research institutes in the world, he jumped at the chance.

Set on fifteen acres of prime real estate on Manhattan's opulent Upper East Side, Rockefeller is a sort of dreamscape. Flagstone walkways lined with London plane trees meander through plantings of Japanese holly, azaleas, pachysandra, winterberry, periwinkle, red maple, and yew trees to a collection of ivied buildings arranged tastefully, like furniture placed just so in a well-designed parlor. There are sculptures by Carmen Teixidor and James Rosati. There are fountains, canopies, and trellises. On a fine day, Nobel laureates play tennis with Nobel aspirants on sun-dappled courts. When it rains they might stroll to Caspry Hall for a chamber music concert, or gather over plates of poached salmon in a dining hall hung with fabulous oils by James Brooks, Joan Mitchell, and Kenzo Okada. It has been said that full professors here are treated like feudal lords. This is of course an exaggeration, but certainly they seem to want for nothing.

John D. Rockefeller believed that scientists deserved and needed a place like this, an oasis of quiet and contemplation, an incubator of talent. As Ron Chernow writes in his biography, *Titan,* Rockefeller "retained a mystic faith that God had given him money for mankind's benefit." Other robber barons would make similar claims to rationalize their greed. But Rockefeller, a devout Baptist, had a pretty clear idea of what God had in mind. His closest advisor, the Baptist minister Frederick Gates, had read William Osler's 1892 textbook, *The Principles and Practice of Medicine,* and been horrified to learn how few bacterial diseases were within the power of physicians to treat. Gates argued convincingly that underwriting a medical research institute aimed at infectious disease would secure Rockefeller, if not a place in heaven, then at least satisfaction on earth.

Gates's timing couldn't have been better, for the final quarter of the nineteenth century was a golden age of microbiology. Louis Pasteur, the son of a humble tanner, had made an ironclad case for the germ theory of infection and disease. Robert Koch, a German bacteriologist, had devised a method for culturing bacteria outside the body that helped him and other scientists make the link between specific bacteria and various diseases. Microbes were traced to maladies at an astonishing rate—tuberculosis, diphtheria, typhoid, pneumonia, gonorrhea, leprosy, syphilis, whooping cough, anthrax, streptococcal, and staphylococcal infections were all tracked down to their sources.

Still, medicine at the turn of the century was not an exact science. There were a growing number of vaccines to prevent disease, but once acquired, few diseases could be cured. The doctor's role was largely to diagnose an illness, explain the diagnosis to the patient and the family, venture a prognosis, offer words of encouragement, and collect a fee. American medical schools were profit-making operations staffed by moonlighting clinicians scrounging pocket change, not by medical scientists interested in uncovering the fundamentals of disease. And medical education was at best an uneven proposition. As described by one analyst, medical schools were "mere groups of local practitioners, nominally, if at all, associated with Universities. . . . Wherever and whenever the roster of untitled practitioners rose above half a dozen, a medical school was likely at any moment to be precipitated." That men (there were few women) of medicine might do research rather than ply their trade was foreign to the American medical establishment. There were medical institutes in Europe—most famously, the Institut Pasteur in Paris—but in the United States the idea of an institution devoted entirely to medical research was controversial, even radical. Gates proposed that Rockefeller endow just such a folly, a place where the best and the brightest would get paid not to doctor, but to think, imagine, and create, where medical scientists could, as Rockefeller later put it, "explore and dream."

In July 1980, Jeffrey Friedman arrived at Rockefeller to pursue his own dream, beginning in the laboratory of Mary Jeanne Kreek. An internist and self-trained psychiatrist, Kreek is by any measure formidable and by most measures indomitable. A graduate of Wellseley College, she was one of only six women in the class of 1962 at Columbia Medical School. Today she is one of only three women who hold full professorships at Rockefeller. The division she heads, the Laboratory of the Biology of Addictive Diseases, spans one and one-half sprawling floors and her office, under strenuous renovation when we meet, seems large enough to shelter a small family. Forceful, quick-witted, and imposing, with a lofty, patrician air, Kreek is unencumbered by modesty, either for herself or for the young doctor she took under her wing twenty years earlier. From the moment she met him, Kreek saw in Friedman's brash audacity a fond reminder of her own.

"Since I was young, I've had this burning sense of inquiry," she says. "Either you ask questions and insist on answering them, or you don't. I insist. Jeff insists."

Kreek had come to Rockefeller in 1964, directly from her medical internship, to work in the laboratory of Vincent Dole. Dole was an expert in metabolic disorders and a pioneer in the use of methadone as a treatment for heroin dependency. When Dole retired, Kreek took on his mission. Like Friedman, she was not drawn to clinical practice; her interest was less in treating drug addiction than in understanding it by illuminating its underlying elements. When Friedman met her, she was searching for chemicals in the brain that, like narcotics, stimulate addictive behavior in rodents and—by implication—in humans.

Observing that some strains of rats self-administered narcotics after just a brief exposure, while others resisted the drugs, it was clear to Kreek that addiction was not merely a "state of mind." She believed that addiction had a physiological underpinning: that it changed the brain in some fundamental fashion, and that a brain prone to addiction was in some way different from other brains. This thinking caught Friedman's attention.

"It occurred to me perhaps for the first time that there was nothing about behavior that was metaphysical; that behavior, in a sense, could be hard-wired in the brain," he says. He'd grasped instantly the concept that so many others had missed—that the loss of control in addiction was not simply a "behavior" (whatever that meant) but that it had a metabolic basis.

"I was starting to ask questions about the role of endorphins in stress and addiction, and he was most intrigued by this," Kreek recalls. She assigned him to study the medical and endocrine effects of long-term methadone treatment in humans by devising a method to measure levels in the brain of beta endorphins, proteins that modulate pain and pleasure signals. It was while puzzling through this that Friedman was introduced by Kreek to another scientist, Bruce Schneider, who was using similar techniques to look for cholecystokinin (CCK), a chemical produced in the gut and the brain known to be important in the regulation of appetite.

Schneider was all too familiar with theories of appetite. He had recently moved to Rockefeller from Mount Sinai School of Medicine, where he'd worked as a postdoctoral fellow in the laboratory of Rosalyn Yalow. Yalow won the Nobel Prize in physiology and medicine for developing (with colleague Solomon Berson) the radioimmunoassay, an ingenious method for measuring barely detectable levels of hormone in the blood. She had used this technique in an elaborate set of experiments to measure CCK levels in mice, and had found that *ob* mice had only about one-fourth as much CCK in their brains as did normal mice. These findings convinced her of what she already believed to be true: that CCK was the "missing link" in *obese* mice. Her paper, published to great fanfare in 1979 in the journal *Science,* stated that *obese* mice "unequivocally" had less CCK than did normal mice, and suggested that a subthreshold level of CCK was the cause of obesity in *ob* mice. Coleman's mysterious satiety factor, it seemed, had been found.

Bruce Schneider knew that Yalow was wrong. He had left her lab in 1977 to join the laboratory of Jules Hirsch, a brilliant research physician at Rockefeller University Hospital who had focused his efforts for decades on the obesity problem, from both a clinical and a research perspective. Schneider had studied CCK for a year and a half in Hirsch's lab, and had come to the firm conclusion that it was not the satiety molecule lacking in *ob* mice. In extensive experiments he had turned up no differences between normal and *ob* mice in either the quantity or quality of CCK. Indeed, he had found that CCK levels varied very little either among mice, or over time in a single mouse. CCK levels didn't budge when mice were fed, and they didn't budge after a seventy-two-hour fast. CCK stayed stable even when mice were overfed and got fat. From this Schneider concluded that while CCK may somehow regulate appetite in both normal and obese mice, it could not be *the* satiety factor absent in *ob* mice. But he wasn't sure what to do with this insight.

"Roz Yalow was a big shot, a member of the National Academy of Sciences, she had just won the Nobel Prize," Schneider recalls. "She was a very self-righteous and imposing individual. And she claimed that she was never wrong. But I was at my wits' end because I kept repeating the experiment and getting the same results. Jules said, 'You've got to write a big paper and report what you've found.' But I was very, very scared."

Schneider tells me this when we meet a few subway stops from his office in Washington, DC. He is an established scientist now, a physician and medical officer for the Food and Drug Administration, and the proud father of grown children, one of them an assistant district attorney in Brooklyn. Still, as he recalls what occurred almost a quarter of a century earlier, his fingers stiffen around his coffee cup, and his eyes narrow. Yalow was a giant in the field of endocrinology, he says, a huge force. For a junior investigator to publicly discredit her work was an act of professional suicide.

"In November 1979 I published my findings in the *Journal of Clinical Investigation*," he says. "One of the nice things about Rockefeller is that people read the literature. They told me, 'Wonderful paper, but boy, are you in shit.' After forty-eight hours I had heard nothing from Yalow, and I knew that she knew I was right. But I also knew that she was merciless. I knew my career was in jeopardy."

Yalow never duplicated her CCK findings, nor did anyone else. Schneider's findings were duplicated many times, by several scientists. Eventually the world was convinced that Schneider was correct—CCK was not the satiety factor, and a mutation in CCK was not at the root of genetic obesity. But CCK remained of great interest to obesity researchers, among them Schneider's new colleague, Jeffrey Friedman, who had grown increasingly intrigued by the idea of molecules affecting behavior. Friedman had sought Schneider's help in designing a radioimmunoassay for beta endorphins, and in turn had assisted Schneider with his CCK experiments. Together they cloned the CCK gene, which was relatively simple to do, given that they knew the protein it manufactured. They found that the gene was expressed—produced protein—in precisely the same way in both normal and *obese* mice. That is, the CCK gene functioned fine in normal mice and *obese* mice, and was clearly not involved in the *ob* mutation. Clinching the matter was that the CCK gene lay on chromosome nine, while the *ob* gene was localized on chromosome six.

"We showed that at both the level of the protein and at the level of the gene, CCK was the same in normal and obese mice," Schneider says. "CCK couldn't be the satiety factor. Yalow never published on CCK again. But she never recanted."

Schneider is a sensitive and emotional man, and he was sorely shaken by what he perceived to be Yalow's betrayal and abandonment. He continued to work with Friedman on CCK, but when Friedman began his search for the obesity gene, Schneider was no longer with him. He had moved on from Rockefeller, to another job and another life. It was a good job, the head of endocrinology at a respected hospital. But as he had feared, his life on the cutting edge of biomedical research was essentially over.

Friedman's, however, had just begun. While Schneider challenged authority and paid dearly for it, Friedman had a knack for endearing himself to powerful men who could help ease his way through the staunchly hierarchical world of biomedical research. At Rockefeller that man was James E. Darnell Jr., a pioneer in the emerging molecular biology revolution who, half a decade later, would be appointed the university's first vice president of academic affairs. Darnell was and is a major figure, the holder of numerous academic awards and prizes, director of a large and influential laboratory, and a gifted mentor to younger scientists. When Friedman met him shortly after coming to Rockefeller, Darnell was studying how cells retrieve information from DNA. Among his many seminal discoveries was that information coming in from outside the cell can turn on and turn off genes in the developing embryo—essentially, that genes are not fixed, but can be manipulated by external forces. This finding would later have profound implications for the understanding of the human genome.

Friedman was trained as a doctor, not as a scientist, and his year in Kreek's lab was meant to be a diverting interlude. In 1981 he was scheduled to enter a fellowship in gastroenterology at Brigham and Women's Hospital in Boston, training that would pretty much assure him an academic post and a profitable private practice. But a year of research had opened his eyes to the new tools of molecular biology, tools whose full power was as yet unknown. Compared to the burgeoning genetics revolution, the fellowship loomed like drudgery. As the end of his research year drew near, he wrote to several scientists, pleading for a job. But without a Ph.D., none would have him. Darnell, sensing the younger man's talent and drive, and impressed by an enthusiastic endorsement from Schneider, took him on as a Ph.D. candidate.

"All my friends were in practice, and here I was, going back for a Ph.D.," Friedman says. "My father [a radiologist] wasn't pleased. He said to me, 'You get a Ph.D., you'll get paid like a Ph.D.,'" meaning not nearly as much as a radiologist. But Friedman sensed that his dad was wrong.

Darnell assigned Friedman to look into the molecular biology of liver regeneration. The liver is among the few organs with the capability to rebuild itself; if as much as two-thirds of a mouse's liver is removed, the organ will more or less re-form within seventy-two hours. Cancer cells have a similar talent, and Friedman began to wonder if there was any connection. "The response to a partial hepatectomy [removal of part of the liver] represents one of the most striking examples of rapid yet controlled cell proliferation that is known," he says. "I was especially interested in assessing whether oncogenes [cancer genes] played a role in normal cellular proliferation, which we now know to be the case. Unfortunately very few oncogenes had been isolated at that point in time, and none of the oncogenes I was testing changed post-hepatectomy."

Darnell recalls that Friedman's interest in CCK and the cloning of disease genes far outweighed his eagerness to explore the nuances of liver growth. What Friedman really wanted, Darnell recalls fondly, was to clone the elusive *ob* gene.

"Cloning was not a trivial exercise back then; there was no cookbook, no kits, no set procedure," Darnell told me. Cloning was particularly difficult when, as in the case of the *ob* gene, there was no gene product to work back from—no protein from which to start. And while researchers had cloned a number of genes from the fruit fly, cloning mammalian genes was a challenge altogether more difficult. "My lab was among the first to understand cloning techniques; we'd been doing it since 1976 or 1977, but we were not at the vanguard of high tech," Darnell says. "It was a very risky business. And cloning a mouse gene that we knew nothing about, like the obese gene, had never been done. You had to use all the techniques as they developed. But when new things come along you depend on the ambitions of young men, and Jeff was certainly that."

"The *ob* gene wasn't CCK, and I wondered, if not CCK, then what could it be?" Friedman says. "I knew the nature of this gene product [the protein] would be interesting. I could envision a strategy for cloning the gene. The revolution in molecular biology was dawning, and this was hard-core, cutting-edge molecular biology. It would take me into the last frontier of science, into neurobiology. I had an instinct that this would be the wave of the future."

Through Bruce Schneider, Friedman had met Rudy Leibel, a scientist and physician who had searched for Coleman's elusive satiety factor and the *ob* gene for several years. Neither Leibel nor anyone else had the vaguest idea what the *ob* gene looked like, or what protein it produced. No one had the slightest notion of its nucleic acid sequence. Like Coleman, Leibel thought that something in the fat cell itself was producing a satiety signal. But what that something was he had not a clue. Leibel was older than Friedman, and had worked with obese and diabetic patients—in particular, children. Unlike Friedman, his interest in this matter was not abstract—he had firsthand knowledge of the pain and suffering these patients endured. He wasn't all that keen to be on the cutting edge of anything, or to ride a wave into the future. What he wanted, quite simply, was to find a cure.

Five

HUNGER

> There is a difference, an important difference, between addiction and compulsion, no matter what the doctors and other innocents may think. Emphasize this. But where does the compulsion come from? From far in the back of the cave where the firelight doesn't reach.
>
> —Hayden Carruth

Rudy Leibel is a small, neatly built man, with hair swept back from a Byronic brow that furrows easily, with irony. An amateur historian and Civil War buff, he has the habit of framing his thoughts in a historical context or a literary one. He has a degree in English literature, and a weakness for poetry—Emily Dickinson and Dylan Thomas being particular favorites. He favors bow ties, and keeps a syringe in his desk drawer to refill his fountain pens with ink. Leibel's restless and expansive intellect makes him something of a curiosity in the fast-paced, high-stakes realm of biomedical research. It was Leibel who recited Frost's "Two Tramps in Mud Time" at Doug Coleman's retirement party.

Leibel was already thirty-eight years old when he came to Rockefeller University in 1979, a few months before Jeff Friedman arrived, a family man whose medical career had until then centered largely around public service and serving patients. He had studied psychiatry in medical school and cultivated a serious interest in psychoanalysis for a while, but eventually decided against it as a career. "I realized that no amount of Freud or Jung was going to bring people back from the brink," he says. He preferred to focus his efforts on helping people avoid the brink entirely, and to that end he decided to specialize in pediatrics and endocrinology. He wanted to get at the heart of the matter, to prevent rather than treat disease. And the biology and genetics of early human development fascinated him. "I wanted to know what inborn errors of metabolism could tell us about the human organism," he says, "what cruel tricks of birth could do to the human form, and especially to the mind." So he took a fellowship in endocrinology at the Massachusetts General Hospital in Boston, and then, after a couple of years tending soldiers and their families on an army base at the height of the Vietnam War, returned to Boston to complete his medical residency at Children's Hospital in Boston. There he met Philip Porter, chair of pediatrics at Cambridge City Hospital and head of child health for the city health department.

Cambridge, Massachusetts, is renowned for its lofty universities, but outside those hallowed halls the city suffers its share of urban woes. There is poverty and its sequelae: shabby housing, drug addiction, child and spousal abuse, and crime. Porter ran a community health care system there: a team of nurse practitioners working out of the city schools that offered guidance on infant care, gave immunizations, and counseled the elderly, battered mothers, and pregnant teens. The nurses spoke Portuguese, Haitian Creole, Spanish, and Mandarin as well as English, and, just as important, they made house calls.

Leibel was deeply impressed by this effort. As vice chairman of pediatrics at the Cambridge City Hospital, he worked on child nutrition and universal immunization and helped found one of the nation's first Special Supplemental Program for Women, Infants, and Children (WIC). Unlike many physicians, he had more than a passing interest in nutrition, for he had witnessed firsthand the impact of poor diet on health. His early research had turned up alarming links between low blood iron levels and learning difficulties in children. That a slight deficiency in a micronutrient could so profoundly affect the brain intrigued him, and made him wonder about other possible links between diet and the brain. Having trained in endocrinology, he had seen more than his share of obese patients in his clinic, and had become increasingly interested in the physiology and etiology of obesity. He had also worked on obese mice in the laboratory, and had seen how they differed fundamentally from ordinary mice. He began to wonder whether there was something amiss in the physiology of obese humans, too—perhaps something, like iron deficiency, that could be remedied. He took the problem more seriously than most, for he had seen what obesity could do to children.

One patient in particular haunted him. Randall, a morbidly obese child of five or six, came to Leibel's clinic on a chilly spring evening in 1977. Leibel performed the usual examination, and unable to find anything physically wrong with the boy, launched into his customary pep talk, blathering on about how Randall could not be blamed for his condition, and that by exercising regularly and cutting back on goodies he could improve his prospects. After two or three minutes of this, the kid's mother shot Leibel a withering look, grabbed her squirming son, and barked, "Let's get out of here, Randall, this doctor doesn't know shit." Watching them leave, Leibel experienced the shameful revelation that Randall's mother was right. Neither he—nor probably anyone else—knew how to treat obesity. He had nothing to offer but pabulum, well-meant yet meaningless clichés. He might as well have prescribed bloodletting or leeches, for all the good it did. Few if any of these kids would get thinner thanks to his advice, and he knew it.

It was while under this Dostoyevskian cloud of self doubt that Leibel happened upon a lecture by Ethan Allen Sims, a physician at the University of Vermont College of Medicine who was then exploring the link between obesity and type 2 diabetes. Sims was interested in whether the metabolic differences observed between fat and thin people were the result or the cause of their body type. Put simply, he wanted to know whether people are born fat or are made fat. He decided that the best way to sort this out was to convince a group of slim volunteers to eat themselves fat and to observe what happened to them when they reduced to their original weight.

Sims was fortunate to have nearby a ready source of experimental subjects: the inmates at Vermont State Prison, sufficient numbers of whom were willing to gorge themselves for science. At first the prisoners proved enthusiastic trenchermen, as much as doubling their usual daily intake of food. But as they fattened, they became increasingly reluctant to overeat. Most found it extremely difficult to gain weight, and eventually some started to drop out of the study. Only twenty made it through the requisite two hundred days, achieving an average weight gain of twenty to twenty-five pounds. Relieved of the high-calorie, low-exercise regimen, all but two of the inmates quickly dropped the newly acquired ballast. The pair of inmates who found it most difficult to lose weight were those who had experienced the least difficulty gaining weight in the first place. It was later discovered that both of these men had a family history of obesity.

From this experiment Sims concluded, as had earlier researchers, that the body was remarkably well equipped to balance energy intake and output, and to reach an energy equilibrium or "homeostasis" at which it felt naturally comfortable. What was particularly interesting was that body weight seemed somehow fixed, and was in most subjects resistant to change over the short term. The prisoners with obesity in their backgrounds were, it seemed, genetically inclined to reach homeostasis at a higher weight than were others; the high-calorie diet only helped manifest their genetic proclivity.

Sims's findings substantiated those of Ancel Keys, an epidemiologist at the University of Minnesota who, decades earlier, had run a study of human starvation. Keys's goal was a noble one: to build a scientific foundation for reviving starving World War II refugees, concentration camp survivors, and prisoners of war. His work was sponsored in part by religious groups, including the American Society of Friends (the Quakers) and the Mennonites, and it had an almost saintly aura about it. Keys selected his thirty-six study subjects from hundreds of applicants, all of them conscientious objectors with a special interest in humanitarian relief efforts. The selected men were healthy and fit and, by modern standards, quite slender—averaging just under 153 pounds on five-foot-ten-inch frames. After an initial orientation, they were put on an austerity diet of whole wheat bread, cabbage, turnips, cereals, and potatoes, with the occasional Lilliputian portion of meat and dairy products. The diet was grueling, meant to mimic as closely as possible that of European famine victims. All but four men in this steel-willed and virtuous lot stuck with the regimen for six increasingly uncomfortable months. Throughout the ordeal, volunteers complained of the cold, even when it was warm. They bundled themselves in sweaters and jackets, and consoled themselves with gallons of steaming black coffee, tea, and water. They lost pound after pound and, not incidentally, their interest in everything but food. (One remarked: "I have no more sexual feeling than a sick oyster.") Even the most refined prep school boys took to licking their fingers, and their plates. They were lethargic, morose, and irritable. To pass the time between scant meals they indulged in increasingly extravagant food fantasies, and many vowed to devote their careers to food production, restaurant work, or agriculture, as if pledging themselves to the priesthood. Most indulged in a sort of "stomach masturbation," compulsively pawing through cookbooks and recipe files like schoolboys fingering a sticky stash of *Playboy* magazines. All in all it was a mortifying experience, aptly summarized in the following tone poem composed by one of the research subjects:

> How does it feel to starve? It is something like this:
>
> I'm hungry, I'm always hungry—not like the hunger that comes when you miss lunch, but a continual cry from the body for food. At times I

> can almost forget about it but there is nothing that can hold my interest for long. The menu never gets monotonous even if it is the same each day or is of poor quality. It is food and all food tastes good. Even dirty crusts of bread in the street look appetizing and I envy the fat pigeons picking at them.

After six months the volunteers dropped, on average, a quarter of their body weight and a somewhat greater proportion of their body fat. In photos they closely resemble the very concentration camp victims whose lives they had hoped to better, all vacuous eyes staring out from hollow sockets in heads that appear far too large. At this point the starvation phase of the experiment was over, and Keys slowly revived the volunteers with a three-month rehabilitation period, gradually increasing the dietary allotment to differing degrees until some were eating as much as six and seven high-calorie meals a day. Free to eat what they liked, some men ate until they vomited, and asked for more. Others ate until they were physically incapable of choking down another forkful, but continued to complain of hunger. Throughout this period they continued to lick their plates, and didn't stop licking until they had regained the bulk of the body fat they had lost in the starvation stage. Regaining their body fat—not their full body weight—seemed to be key to regaining their composure. This hinted that there might be a mechanism, possibly in fat tissue, that was "telling" their too-thin bodies to refuel.

Together, the Sims and Keys studies gave evidence of a theory of weight stabilization that had been bandied about for nearly a century: that weight is controlled by a sort of fat thermostat—Gordon Kennedy's "lipostat"—that senses how much fat there is on the body and adjusts eating and energy expenditure accordingly to maintain a steady state or "set-point."

After hearing Sims's talk, Leibel became convinced of the merits of this theory, and of the biological determinants of body weight regulation. He grew absorbed with the idea of finding the missing link—the signal that tells the body to eat, or not. "I became convinced that there was a signal from the body fat to the brain that would protect fat stores," Leibel says. "There would have to be to support growth and fertility." Leibel recalled the work of Rose Frisch, a Harvard researcher who had published extensively on the connection between body fat and fertility. Frisch had shown in studies of female athletes that women require a minimum level of body fat—at least 12 percent of total body weight—to maintain their fertility. Leibel guessed that some signal emitted by adipose tissue controlled both body weight and fertility. This certainly made sense in the evolutionary context: painfully underweight women are unlikely to have the strength to effectively feed and care for themselves, let alone bear and nurse children. Emaciated men are ill-prepared to gather food or hunt or defend their families. The shutting down of fertility is an adaptive mechanism to protect the ultrathin from nonessential energy expenditure, and to prevent the unfit from generating more hungry mouths to feed. Why this same mechanism might "tell" the body to also lose gained weight, as seemed to be the case for most men in the Sims study, was less clear, but Leibel was determined to find out. And he knew just the man to help him.

Rockefeller University clinical researcher Jules Hirsch was a highly regarded champion of the set-point theory and among its most eloquent proponents. He ran a clinic at Rockefeller Hospital in which he studied feeding behavior in obese and normal humans, and also did animal studies. Through a series of ingeniously designed experiments in rodents, he had shown that mammals are genetically "set" to sprout a certain number of fat cells, but that this number could be "reset" by manipulating diet at different points in the animal's development.

Human infants are born with an estimated 5 billion fat cells, one-fifth or so of the average adult allotment. Individual fat cells (adipocytes) vary wildly in size, depending primarily on the amount of stored triglyceride they contain. When the amount of energy taken in chronically exceeds the amount of energy one needs for daily living, adipose tissue grows, by an increase in either the size or the number of fat cells, or both.

In early infancy, fat cells fill like balloons with triglyceride, and their number remains fairly constant. At about three months, the number of fat cells begins to increase. Fat cell proliferation continues through adolescence and then starts to slow, but contrary to earlier theories, new fat cells can and do sprout and die throughout life. Fat cells are remarkably malleable, but even the most accommodating of them has its limits, and when existing cells pump too full with lipid, new fat cells grow to shoulder the load. Mild obesity usually reflects a plumping up of individual fat cells, while more severe obesity, and obesity arising from childhood, is usually manifested by a substantial increase in the number of fat cells: average-weight adults have 25 to 30 billion fat cells, while the obese may have as many as 200 billion.

Hirsch noticed that genetically slim rats fed an exceedingly rich diet early in life grew an usually large number of fat cells, and that this fat was difficult to shed. This was not entirely surprising: all cells stubbornly resist obliteration. This helped explain why dieters in his obesity clinic at Rockefeller Hospital so often hit the "wall"—shrinking fat cells fought back, sending ever more frantic "eat" signals to their brains. Eventually, as Keys's starvation experiment had shown with frightening clarity, these signals all but drown out other messages to the brain—and become almost impossible to ignore. A reduced obese person is not the same as another person of the same weight who has never been obese. Just as Leibel suspected, their brains are different.

Leibel found in Hirsch a kindred spirit. In 1978, while in New York to attend a pediatric meeting, he decided to make an impromptu visit to the older scientist's laboratory. Leibel recalls leaving his hotel and walking across Central Park toward Rockefeller with a growing sense of urgency, only to find when he arrived that Hirsch was out of town. Undeterred, he corralled several members of Hirsch's laboratory into a bull session that ran through the afternoon and into the evening. "I met several people who believed as did I that there was something in body fat communicating with the brain, but of course, no one knew what," Leibel says. "I thought, okay, enough talk, I'm going to do this." Leibel returned home and arranged a leave of absence from Cambridge City Hospital and from Harvard Medical School, where he was an assistant professor. He and his wife Lulu put their rambling Victorian on the market, sold most of the family's possessions, packed up their two young daughters and moved into an eight-hundred-square-foot apartment in Rockefeller faculty housing. "I took what amounted to a postdoc appointment at half my former salary," he says. "But there was no way around it. I needed to know."

What Leibel needed to know was how human eating behavior was dictated by human physiology and, ultimately, how genes orchestrated this process. This was thorny, contentious stuff. The concept that human behavior could be influenced by genes was, on many levels and to many people, repugnant. In the United States individual potential and prerogative are held supreme, and the idea that something as fundamental as body weight regulation was biologically determined grated painfully against American notions of free will and self-determination. John Locke, the seventeenth-century philosopher and an early adherent of British empiricism, famously argued that humans are born into this world "blank slates" to be written upon by experience. Americans, at least in theory, are very fond of this idea, one that seems to disqualify "nature" as a player in the nature/nurture debate. But even Locke didn't go that far. In 1692 he wrote: "I confess, there are some men's constitutions of body and mind so vigorous, and well fram'd by nature, that they need not much assistance from others; but by the strength of their natural genius, they are from their cradles carried towards what is excellent; and by the privilege of their happy constitutions, are able to do wonders." Later in the same essay he observed: "Some men by the unalterable frame of their constitutions, are stout, others timorous, some confident, others modest, tractable, or obstinate, curious or careless, quick or slow." It seems that even

Locke believed that while we are born with the free will to choose our leaders, not even the strongest among us are able to shape their own characters, or hone their own body types. After all, there was instinct to consider, the difficult-to-describe quality by which animals—even human ones—seem driven from birth to behave in a certain manner.

In the mid-twentieth century, animal behaviorists Karl von Frisch, Konrad Lorenz, and Niko Tinbergen argued that instinct could be broken down into what psychologist B. F. Skinner called "atoms of behavior," simple routines and subroutines that were hard-wired into the organism, and therefore out of mindful control. But as the century progressed, there was a growing resistance to applying these or similar rubrics to the study of *human* behavior. Social scientists in particular grew contemptuous of theories that "biologized" people in this fashion. When, in 1975, Harvard biologist E. O. Wilson published his groundbreaking treatise, *Sociobiology,* on the genetic correlates to behavior, he was decried by a phalanx of critics who argued that while the instincts of animals might well be inborn to some extent, the behavior of humans was shaped almost entirely by culture.

Wilson's critics were not wackos—among them were several of his most eminent Harvard colleagues, and they had good reason to be alarmed. The claim that humans are endowed from birth with certain immutable characteristics—among them intelligence and criminality—has underpinned some of history's most heinous episodes. Ideologues have appropriated and contorted this argument to rationalize racism, sexism, xenophobia, and eugenics. As author and critic Arthur Koestler wrote: "The attempt to reduce the complex activities of man to the hypothetical 'atoms of behavior' found in lower mammals produced next to nothing that is relevant—just as the chemical analysis of bricks and mortar will tell you next to nothing about the architecture of a building." There was, it seemed, something vulgar about reducing human behavior to the realm of the merely physical.

Obesity researchers of the time strove to distance themselves from this rancor, and worked hard to remain on the politically correct side of the argument, clinging to the idea that the decision to eat was shaped entirely by psychological factors amenable to psychological treatment. Leibel had witnessed the nature/nurture upheaval firsthand in Cambridge, and he wanted no part of it. Politically liberal, he was a firm believer in the power of humans to shape their own destiny, regardless of genetic heritage. Still, his experience as a clinician told him that, sacrilege or no, human eating behavior was to some extent biologically driven, just as were physiological variables, such as blood pressure and heart rate. He found the public's reluctance to accept biology as a factor in obesity enormously frustrating. "When it came to obesity, people kept putting forth these vitalist arguments," he says. Vitalism is a nineteenth-century theory, the basic tenet of which is that the living differ from the nonliving in that they possess an intangible inner energy, a quality that is difficult to describe and impossible to quantify. (It might fairly be compared with the Middle Ages preoccupation with the "philosopher's stone," which Carl Jung wrote was considered by alchemists to be "the mystical experience of God within one's own soul.") Vitalism held sway in Europe for about a hundred years, until being called into question by German chemist Friedrich Wohler's accidental synthesis of urea, an organic molecule. Clearly if organic molecules—the molecules of living things—could be cooked up by man, there was nothing "vital" or otherworldly about them.

Still, the legacy of vitalism lived on in the new alchemy of psychology, which sometimes offered explanations for human behavior that seemed to verge on the mystical. For many, to say that something was "psychological" was to say in effect that it was outside of biological control. Too often psychologists did not pause to consider what was prompting these "psychological" effects. Schizophrenia, for example, was blamed on the neurotic behavior of one's mother, not on an unruly clash of brain chemistry. And overeating was explained as the sad consequence of some vague oral fixation. Leibel was interested in something more fundamental. Psychological forces, he reasoned, must be propelled by tangible phenomena, not by some metaphysical force brought on by a vague childhood trauma. He did not question that psychology played a role in whether or not people chose to behave in a certain way, but what interested him were the forces that drove that psychology. Finding the mysterious factor that drove eating, Leibel sensed, would go a long way toward putting vitalist theories of eating behavior to rest. Genetics, he felt certain, played an important role in determining body weight regulation in humans, just as it did in mice.

The *ob* mouse offered a dramatic example of biologically driven eating, one in which the presence of a single gene directly dictated a behavior. But genetics also seemed to underlie less extreme differences in weight regulation. For example, whereas some altogether normal rodent strains got fat when offered a plentiful "cafeteria"-style diet, other equally normal strains did not. It hardly seemed rational to suggest that the greedier rodents were "orally fixated," or driven to overeating by an overbearing mother. Clearly, genetic factors in the different rodent strains were important for weight regulation. And there was growing circumstantial evidence to back the genetic theory in humans as well. Studies of adopted children had begun to hint at the significance of genetics in human obesity.

Identical twins tend to have remarkably similar body mass indexes, far more similar than do siblings or even fraternal twins. This is true even for identical twins separated at birth and raised far apart in different adoptive families. The BMI of adopted children usually correlates nicely with that of their biological parents, but not with that of their adoptive parents. This strongly suggests that genetics or congenital effects, not "psychology," play the larger role in human obesity.

When Leibel was getting his start in Hirsch's lab in the late 1970s, all this was known. But there was no smoking gun, no physical evidence of a genetic link to obesity in humans, because the molecular tools to find such a link did not yet exist. Leibel had spent several years in informal collaboration with Doug Coleman, who often came to New York from Bar Harbor to visit Hirsch's lab. With Coleman's guidance, Leibel began a painstaking search for the elusive satiety factor in the adipose tissue of normal mice.

Leibel had worked with *ob* mice since his training years in Boston, and his experience with organic chemistry stretched back several decades—to summer days working in his father's dry cleaning business. "The organic solvent extractions involved in looking at the biochemistry of adipose tissue were similar to those involved in dry cleaning," he says. Leibel, like Coleman, assumed that the satiety signal was tied up with glycerol, a component of the triglyceride molecule. Blood glycerol levels correlate with the number of fat cells in the body, so it made sense that high glycerol levels might signal the body to stop eating. To test this assumption, Leibel injected mice with glycerol. The mice reduced their eating for a while, but after about a week, their appetites returned. Leibel repeated the experiment in humans, who responded not a whit. Clearly, glycerol was not the answer.

Over the course of a few years Leibel refined his search strategy. The facts were these: Neither he nor Coleman nor anyone else knew what the satiety factor was, let alone what produced it. What they did know was that the absence of such a signal or the inability to respond to it caused *ob* and *db* mice to eat uncontrollably. Their working hypothesis was that a mutation in the *ob* gene produced mice that were unable to manufacture a working satiety-signaling protein, and that the *db* mutation produced mice that had the protein but lacked the ability to detect the signal. They agreed that the best way to locate the *ob* protein was to work back to its source, the *ob* gene, the mutation that had alerted Coleman to the existence of the satiety factor in the first place.

"I believed that there was only one way to break through a growing morass of poorly integrated phenomenology in this field," Leibel says. "We had to reenter it through a specific molecular mechanism for the control of body weight as exemplified by these mutant animals. We would have to clone the *ob* and *db* genes based upon their respective positions in the genome."

Jeff Friedman was at the time working out of a small corner of Jim Darnell's well-appointed lab, mastering the latest techniques in molecular biology. He was by then an assistant professor, in his early thirties and living alone in Rockefeller housing. Leibel says he approached Friedman with the *ob* cloning proposal in the spring of 1986. Friedman recalls this episode differently, insisting that it was he who approached Leibel, well after he had hatched his own strategy to clone *ob* and begun planning its execution. The passage of time clouds memory, and it is doubtful that this conflict will ever be resolved to the satisfaction of both scientists. But what is certain is that their collaboration was fortuitous. Leibel understood the physiology of obesity, the cell physiology, and had thought through the underlying scientific problem. Friedman had the tools of molecular biology.

Both men knew the project was a long shot, an all-or-nothing proposition. Either they found the *ob* gene or they didn't; no gene, no glory. This was dangerous for Leibel, an associate professor at Rockefeller whose career was well established. But for the younger and more junior Friedman, it meant flirting with professional annihilation. "He knew that if it didn't work, he would barely get a paper out of it," says his brother, Scott Friedman. "He was taking a huge risk, a risk most people, including myself, would not have taken."

Locating a gene in a mammalian genome without knowing its protein product is something like finding the home of a reclusive uncle who lists his address as "Someplace, USA." Before even thinking about cloning the gene, the scientists had to home in on its general location, to distinguish the rough stretch of DNA on which the gene lay from the rest of the chromosome. For this they would need a map.

Genetic maps, like all maps, are created by locating points relative to one another. When we think of, say, Paris, France, we describe its location as relative to other things, as south of Normandy, north of Lyon, so many miles from the Atlantic Ocean. Genetic maps are similar, designed by positioning genes in relation to one another on a chromosome. The very existence of genetic maps is made possible by the way sexually reproducing species pass down their genes. The genetic material in higher organisms is arranged in genes, copies of which are passed down from parent to child. Gregor Mendel had assumed that all genes passed down independently of each other, gene by gene, but Mendel was only partially correct. Some genes are passed down together, in strands.

This transforming discovery of genetic linkage was made in the early twentieth century by Thomas Hunt Morgan, the "father of American genetics," with the help of his brilliant student, nineteen-year-old Alfred Sturtevant. Morgan and Sturtevant studied fruit flies, and found that certain fruit fly traits were passed down in a kind of package deal from parent to child. The closer one gene was to another on a chromosome, the more likely the two genes would be passed on together. Genes located on different chromosomes, or at positions far apart from each other on the same chromosome, were passed on independently of each other, just as Mendel predicted. Morgan observed which traits were likely to be inherited together and how great that probability would be, and Sturtevant used that information to construct the first genetic map. Sturtevant's map charted the links among wing shape, eye color, and body color of fruit flies, traits that he observed were generally inherited together. The fruit fly was the example that proved the rule: mice and men and all living things also tend to inherit genes in this way. Sturtevant's rough genetic map led to research that underlay the human genetic map we have today, the map that charts the genes in the human genome.

A complete human genetic map was more than a decade away when Leibel and Friedman started their hunt for the *ob* gene in 1986. Nonetheless, some disease genes had been laboriously hunted—and even located—in humans. A team in Boston had bagged the gene for a type of muscular dystrophy. And Leibel had worked personally with Nancy Wexler, the charismatic Columbia University researcher whose studies of Huntington's disease in a particularly susceptible population on the coast of Venezuela would later lead to the discovery of the Huntington's gene.

These advances were good news, and they made the discovery of a disease gene, such as *ob,* marginally more likely. Earlier researchers had located a number of "markers," DNA sequences that serve as signposts along the twisting length of a chromosome, and from which biologists can "walk" metaphorically toward their genes of interest. Unfortunately most of these marker sightings were in humans, not mice. The Rockefeller team would have to find most mouse markers themselves through the laborious process of crossbreeding laboratory animals one pair at a time, and noting which physical traits were inherited together. As Morgan and Sturtevant had found, traits that are inherited together are encoded by genes that are close together on a chromosome. By starting with genes that were already mapped and then working back, Leibel and Friedman hoped to rough out a map that would help them locate the *ob* gene, and its molecular doppelgänger, the *db* gene.

Leibel scraped together the money from Rockefeller and other grants to hire two young scientists to work on the project: Streamson C. Chua, a recent M.D./Ph.D. graduate from Columbia University, and Nathan Bahary, a medical student from Cornell University. Chua had worked on a cow gene related to a human gene, but had no experience with making molecular maps in mice. Bahary was smart and ebullient, but green. "We continued to get a lot of advice from Doug Coleman, with whom I was in constant contact," Leibel says. "But there were many things everyone had to learn de novo."

Friedman and Leibel tracked the *ob* and *db* genes in thousands of animals, and prepared increasingly more refined genetic maps. But after a year they realized that there were simply not enough markers to guide them. They desperately needed a way to shortcut the process. Chromosome microdissection, a way to create markers by painstakingly dividing a chromosome into several fragments, looked like a promising solution. Only a handful of people in the world could manage this intricate technique, and none of them were in New York. So in 1987, graduate student Nathan Bahary was dispatched to London to learn microdissection in the laboratory of biologist Steven Brown.

Brown's laboratory at London's St. Mary's Hospital had pioneered microdissection just a few years earlier. "It was an exciting time because the promise of positional cloning was everpresent and there was much excitement about the information that would emerge from cloning some of the most interesting mouse mutants," recalls Brown, now director of the MRC Mammalian Genetics Unit and UK Mouse Genome Centre in Harwell, England. Microdissection involved chopping up chromosomes very precisely, an absurdly difficult process.

Bahary built his experiments from the ground up, crafting even his own glassware, sets of exquisitely fine needles and pipettes. He grew mouse cells in culture, swelled them with saline solution, and dropped them from a height of two to six feet onto a microscope slide, where they burst and spilled their chromosomes. He then placed the slide upside down on the microscope, so that he was looking through the glass slide itself to see the chromosomes, suspended in a droplet of liquid that dangled from the bottom of the slide. This unconventional—and precarious—positioning of the slide was necessary to allow him to reach the chromosomes with his tiny cutting tools.

Bahary spent four months in Brown's lab, sweating out the intricacies of this impossibly complex maneuver in a subbasement laboratory. It was a damp and dismal place, perched directly above the Paddington subway line. "Every time you heard a rumble, you had to grab everything and hold it down, or lose it," Bahary recalls. Not surprisingly, he regularly lost several days' work to the rumblings of the train. With no money to pay for a hotel room, he slept on a cot in a lecture hall, and showered in the washroom at the hospital pool. He spent days without encountering a photon of sunlight. Yet through it all he maintained an irrepressible enthusiasm. He was venturing where few had gone, and he knew it.

"His microdissection was a work of art," Leibel says. "It was a technical tour de force, just beautiful."

Bahary returned to Rockefeller and over a period of several months used the technique to create molecular clones that were mapped back on the mouse crosses for *ob* and *db,* further reducing the interval of contiguous DNA. The team reported the initial mapping of the *ob* gene in a series of papers published in the early 1990s. It was a thrilling time, but also exhausting and highly charged. As the probability of cloning the gene grew, tensions in the lab curdled into hostility. Never known for his grace under fire, Friedman had repeated blowups with laboratory personnel. Streamson quit the project, as did Don Seigel, another postdoctoral candidate, and at least one laboratory technician. "Jeff turned this into the *Apocalypse Now* of molecular biology," recalls Siegel, now an instructor at Albert Einstein College of Medicine. It is perhaps no coincidence that *Heart of Darkness,* the novel upon which *Apocalypse Now* is based, is Friedman's avowed favorite.

Friedman was worried that Leibel and his mentor, Jules Hirsch, would steal credit for the discovery. His suspicions grew, and eventually he demanded that Leibel physically distance himself from the project, essentially staying out of the lab where *ob* was being cloned. "Because of my reputation in the field of obesity research—which had helped us obtain National Institutes of Health funding in the first place—he stated that he felt I was getting an inordinate share of whatever publicity surrounded this project," Leibel says. Leibel was the senior of the two, and he felt a responsibility to protect the project. In early 1993 he decided that to avoid further conflict he would step away from day-to-day involvement. But he remained in close contact with Friedman and his staff, who continued to seek his counsel and advice.

By the time Leibel officially stepped back, the team had spent six years narrowing the location of the *ob* gene to between two markers a few hundred thousand base pairs apart. This was close, but not close enough to home in on the gene. About seven months earlier, in May 1992, Friedman had hired as a postdoctoral fellow Yiying Zhang, a freshly minted Ph.D. from New York University Medical School. Zhang came to Rockefeller two months after the birth of her first child, prepared to make history. "People said that I was crazy trying to use positional cloning to find a gene," she says. "Only a few projects like this had been successful. But the idea was very sound, and I figured we had a good chance."

Zhang employed a technique that Don Seigel had also used—yeast artificial chromosome (YAC)—to clone the potential *ob* region onto another piece of DNA, where it could be more easily manipulated. A laboratory technician, Ricardo Proenca, narrowed the field further using another technique called exon trapping. Exons are the sections of a gene that code for a specific protein. By snatching these sections of DNA away from the introns, the sections that do not code for a protein, Proenca reduced the number of gene candidates to about two hundred. Zhang and Friedman then narrowed the field to four genes that lay inside the area where they expected *ob* to lie.

"We knew that *ob* resided in the region, and we fished out every possible expressed gene," Zhang says. This expression is evidenced by the presence of messenger RNA. mRNA is a molecule closely related to DNA that transcribes and transports a copy of the DNA stripped of its introns. This copy becomes a template for making a protein. In the case of *ob,* the normal protein is longer than the mutated protein, which terminates prematurely and results in the *ob* phenotype or body type. Searching through the morass of expressed genes, Zhang found one whose DNA contained a premature stop signal, indicating that it was a mutant. What was particularly telling was that she found this gene in a fat cell. "We sequenced the whole gene, and we found the secretory protein for which it coded, a small molecule that functioned as a hormone," she says.

The protein had all the makings of Coleman's mysterious satiety factor: a molecule produced in the fat coded for by a gene that was mutant in *ob* mice. But there was no telling whether it truly was the *ob* gene until the hypothesis was tested in animals. By then the team had determined that there were actually two versions of the mutation: *ob2j,* in which there was no production of satiety factor, and *ob,* in which there was an overproduction of the impotent, truncated form of the factor. Either mutation resulted in an *ob* mouse. The next step was to find whether the mRNA from the *ob* gene was present in either of the two types of mutant mice—that is, to see if the *ob* gene was actually expressed in mutant animals.

"On Friday, May 6, I set up an experiment on the normal mouse gene to compare it with the mutant animals," Zhang says. "The next day, Saturday, I brought my daughter, Irene, into the lab." It turned out that the *ob2j* mutant mice didn't produce any mRNA at all, and that the *ob* mutant mouse did. "At that time I was 90 percent sure we had the gene. I ran to Jeff and told him, and he said, 'Irene brought good luck.'"

One final test was needed to confirm the finding: a Northern blot analysis to quantify the amount of mRNA produced by the *ob* mutant. That experiment, assigned to another postdoctoral fellow, Margherita Maffei, was only half completed. Maffei was at a friend's wedding that weekend and out of touch, but Friedman couldn't wait for Monday. "I called her home and left a message," he says. "I went through her stuff, and found what I needed and finished setting up the experiment." At six o'clock the next morning, Friedman was back for a look.

The fat cells from the *ob* mutant produced far more than normal amounts of mRNA, but when the mRNA translated into protein it produced a short, nonfunctional version of the protein that was immediately degraded. The *ob2j* cells produced no protein at all. Seeing this, Friedman knew he was golden. His team had shown that two different mutations in the *ob* gene had affected levels of an adipocyte-specific mRNA encoded by the gene they had mapped to the location where the *ob* gene was predicted to be. "I looked and there it was, right there on the gel," he says. "That moment was the closest thing to a religious experience I've ever had."

Friedman called his future wife, Lily Safani, to break the news. Then he called his mentor, Jim Darnell, to thank him for his support. He did not call Rudy Leibel, Nathan Bahary, Doug Coleman, or the other scientists who had devoted years of their lives to the project. He decided it was better to share the news with them later. He had other things on his mind.

Friedman is a serious sports fan, the sort of fan who remembers the really big plays five or ten years after they happen. That Sunday the Knicks had a playoff game against the team's arch rival, the Chicago Bulls. New York had never won a playoff against the Bulls, and Friedman and his cousins had third-row seats. (These particular cousins owned a profitable Weight Watchers franchise in New Jersey.) Friedman recalls the game vividly, telling me six years later that guard John Starks had a "phenomenal dunk" against Michael Jordan, and Starks's breakaway layup pulled the team ahead with fewer than forty-four seconds on the clock. After the game, Friedman invited Lily and a pal to share a bottle of celebratory champagne at Pete's Tavern, the fabled saloon where the writer O. Henry is said to have penned his most famous short story, "The Gift of the Magi." In the story, a man sells his pocket watch to buy his lover an elegant comb for Christmas, not knowing that she has sold her hair to a wig maker to buy an extravagant watch chain for him. But this touching tragicomedy of errors was unlikely to have played a role in Friedman's thoughts as he toasted his triumph that night.

Friedman called Leibel the next morning and invited him to the lab to confirm the findings. Leibel was overjoyed, but promised to keep the secret until the gene was sequenced and the results published. Friedman returned to the lab to work through every contingency, filled with dread that another group would beat him to publication. "There was an abstract published by someone looking for the same gene, and there was a young scientist in Japan looking for it," he says. There was also a group in San Francisco and a group in Seattle, and probably others that he hadn't heard of, including, he later learned, a team in Boston. None of them was close, but at the time Friedman couldn't have known that, just as his competitors could not have known that he had already found the gene. "We kept this all under wraps," Friedman says, adding that his team spent the next six months learning everything they could about the *ob* gene and its mutations. They charted the exact sequence of base pairs, and found that the *ob* gene coded for a protein of 167 amino acids, the *ob* protein, and located its human counterpart, a slightly different version they called *Ob*. The *ob2j* mutation turned out to be the result of a change in a single base pair: a simple trade of thymine for cytosine. It was this tiny glitch that transformed what would otherwise have been a normal mouse into the lumpish *ob*.

Friedman shared the news with Doug Coleman that fall, when the older scientist came to Rockefeller on one of his regular consulting visits. Friedman, Leibel, and Coleman were on their way to dinner at a German restaurant, and in excellent spirits. "It seemed to me that Jeff was walking on air," Coleman says. "He told me he'd found *ob*, and then he swore us to secrecy, because he was worried about the patent. I thought that was really strange, because it occurred to me that the information should be shared with everyone, not kept secret. It occurred to me that if this could help people, we should let the world know."

"Positional Cloning of the Mouse *obese* Gene and Its Human Homologue" appeared as a cover story in the journal *Nature* on December 1, 1994, precisely one day after Friedman had filed for a patent. The cover shot shows a set of scales on which an *obese* mouse easily nails a balance counterweighted with a pair of normal mice. Friedman is listed as senior author, Yiying Zhang as first author, and Proenca, Maffei, and two laboratory technicians are named as coauthors. Leibel and Bahary were acknowledged in the fine print at the end of the paper as "important contributors to the early phases of this work," as was Lily, Friedman's fiancée, who was not a scientist.

Leibel was flabbergasted. He had done the map work, drummed up much of the grant money, served as a guiding intelligence on the project, and, with Nathan Bahary, pretty much held things together for half a decade. "I of course assumed my name and Nathan's name would be on the paper," he says. "I was terribly shocked, and hurt, to find that they were not." Leibel and Bahary believe that Friedman assigned them the vague auxiliary status of "early contributor" deliberately, to guarantee their getting cut out of credit—and potential financial reward—for the discovery.

"It was like coauthoring the first eleven chapters of a book, then being left off as coauthor because you didn't write chapter twelve," Bahary says. Bahary says he could only then understand why Friedman insisted on the separate publication in 1993 of the microdissection results. Friedman had taken the unorthodox step of writing the microdissection paper himself, crediting Bahary as first author, and, without Bahary's permission, submitting it to a minor journal. This was, to put it kindly, unconventional behavior, and at the time Bahary found it troubling. "I couldn't figure out why he wanted to publish those results in a separate paper when we were so close to finding the gene," Bahary says. "He told me he was doing it for me, but in retrospect the only logical reason I can think of for his writing that paper was the desire to later leave me off the *Nature* paper." By publishing Bahary's virtuoso feat early, Friedman was technically free to omit a detailed description of it from the *Nature* paper, and thereby avoid sharing credit with Bahary for the *ob* discovery.

Zhang, too, is adamant that Leibel and Bahary deserved to be named as codiscoverers. She says that Friedman attempted to distance even her from the narrative of the gene's discovery in order to minimize her claim. She says he took away her final experiment—the one to which all her previous work had led—to fragment credit for the discovery so that he would be the one scientist left with longevity on the project. "Once we cloned the gene Jeff turned around 180 degrees," she says. "From his point of view it was as if I was not there. After you clone a gene, there is a lot you can do, and I had an experiment in mind. Jeff had promised me when I started that I would be able to do this. But he did not keep his promise—he took the whole region away from me and gave it to a medical student who had just gotten to the lab." This was Jeffrey Halaas, an M.D./Ph.D. student who had signed onto Friedman's lab with the goal of finding the receptor for the *ob* protein. Halaas intended to locate the *ob* receptor in the mouse genome, and to use this discovery as the basis for his doctoral thesis. But his more immediate task was the one that Zhang had hoped to perform—testing the newly discovered product of the *obese* gene to prove conclusively that the satiety factor was a hormone that regulates body weight in living animals.

The Nobel prize-winning biologist Roger Guillemin once told Friedman at a meeting that in a rational world, the *obese* gene would be renamed *lepto,* from the Greek word for "thin." It might be considered a gene for thinness, in the sense that mice grow fat without a properly working copy. Friedman picked up on Guillemin's suggestion and christened the *ob* gene product "leptin." The great hope, of course, was that leptin would work as an appetite inhibitor in humans. The first step was to test it in mice.

Even normal mice produce only minuscule amounts of leptin, and to test its potency, the investigators needed a substantial supply. Friedman called on Stephen Burley, another Howard Hughes Medical Institute investigator at Rockefeller, to churn out more of the hormone in *E. coli* bacteria. The challenges were formidable. To be viable, proteins like leptin rely not only on a chemical structure, but also on a certain topology: they must be folded in a particular manner. *E. coli* bacteria produce only an inactive (unfolded or improperly folded) version of leptin. After purifying this inactive form, Burley's lab had to find a way to refold the leptin protein to render it active. It took two skilled people months to work out the solution to this refolding problem, and to produce the very large amount of active material necessary for animal experiments.

Halaas injected the newly synthesized leptin into *ob, db,* and normal mice. To his delight, the protein worked just as Coleman's theory had predicted. *db* mice were not in the least affected by the leptin injections, *ob* mice slimmed down dramatically, and normal mice, pumped with an excess of leptin, lost every gram of their body fat. Leptin, then, was the long-sought satiety factor. And the marvelous thing was that unlike starved mice or starved humans, the leptin-pumped mice were healthy and muscular. Their fat was gone but their muscle mass was spared. Soon thereafter, several other laboratories tested leptin and reported similar results. Leptin, it seemed, was not only Coleman's mysterious satiety factor—it was a fat person's dream incarnate. Friedman had struck gold.

Leptin was an overnight sensation. Rockefeller's switchboard ignited with callers frantic to participate in human trials of the hormone. Food writer and columnist Jeffrey Steingarten spoke for weight watchers the world over when he told a scientist he'd be willing to inject the stuff "into my eyeball," if necessary.

Friedman was prepared for this roar of enthusiasm; he had spent eight years tracking down the molecule, and he fully understood its value. Long before publishing the findings, he had consulted with a number of financial experts to hammer out the terms for licensing the gene to an outside company. Among these was Rockefeller University general counsel William Greiser. "I started getting calls from various companies before the *Nature* article—but frankly I had no idea of its value," Greiser says. "Jeff told me it was valuable, but how do you know whether it's one million or ten or twenty? The difference in this case was that I wasn't getting calls from the usual technology transfer people. I was getting calls from company presidents and CEOs, who all wanted to be the first one there."

To sort through this morass, Greiser took what, at the time, was the extraordinary step of conducting a silent auction, allowing interested companies to bid in writing, and calling in the highest bidders for independent negotiations. "Our thought was, which company would be the best to carry this forward?" he says. Millennium Pharmaceuticals was an obvious choice: the company had a long-standing interest in obesity, and Friedman was a founding member of its scientific advisory board. Millennium CEO Mark Levin called Greiser several times, assuming that he had a lock on the deal. "But you see," Greiser says dryly, "it wasn't Jeff's decision to make."

Apparently it was Rockefeller's decision, and Rockefeller chose another company. After months of arduous negotiations, drug giant Amgen of Thousand Oaks, California, wrestled down the rights to leptin for an astonishing up-front $20 million payment, reportedly the largest ever for a university-held patent. (The day the deal closed, Amgen's stock price spiked 5.5 percent.) Amgen agreed to make future payments to Rockefeller as leptin passed certain "milestones" toward what the company hoped would be its approval as a breakthrough weight loss drug. Greiser is vague about these milestone payments, saying only that they amount to several times the initial $20 million, and that at least one milestone had already been reached. If all went well and the drug went to market, the payment would hover in the $100 million range, not including royalties on the sale of any drugs developed through its application.

One-third of the first Amgen payment went to Rockefeller, and one third to the Howard Hughes Medical Institute, the nonprofit medical philanthropy that underwrites Friedman's salary. The remaining chunk—about $7 million—was distributed to the scientific team. Friedman won't say what his share was, but it is estimated to be between $5 and $6 million. He is entitled to many more millions with each passing milestone. Contrary to his father's prediction, Jeff, the wise-guy younger son, had proved that he could make a decent living in academic medicine after all.

Bahary, Leibel, and other critical contributors to the leptin find were not so lucky. Because they were not directly involved with the project at the "moment of discovery," the lawyers determined that they could be cut out of the deal. Doug Coleman protested. "I was chairman of the study section that gave Leibel and Friedman their first NIH grant, and their grant renewals," he says. "You couldn't disassociate either one from the project. They were both critical." David Luck, a cell biologist who was then vice president for scientific affairs at Rockefeller, agreed in principle. But Luck's first loyalty was to Rockefeller, and Jeff Friedman was not only the protégé of some of the university's most powerful figures, but also a Howard Hughes investigator who had brought a good deal of money to the university. Given Friedman's refusal to share credit for the finding, there was really nothing Luck or anyone else could do.

Bahary and Don Seigel each received a modest lump sum under the condition that they seek no further compensation. Leibel was offered a somewhat larger amount and similar terms. He at first refused, but after some soul-searching and prodding by his wife, he grudgingly gave in. "I could either fight this thing with a lawsuit, or go on with my life," he says. "I decided on the latter."

Leibel left Rockefeller and set up his own laboratory at Columbia University, where he is professor of pediatrics and medicine and head of the Division of Molecular Genetics. Zhang and Streamson Chua have since joined him, and they have together made great strides in leptin research.

Friedman was named full professor and head of laboratory at Rockefeller in the summer of 1995. A year later he married Lily Safani, of the famed Safani Gallery family, and two years after that, at the age of forty-four, he became the father of twin daughters. On May 1, 2001, he was elected to the elite National Academy of Sciences, one of the highest honors that can be accorded any U.S. scientist or engineer.

The man behind it all, Doug Coleman, was also inaugurated into the National Academy, in 1999. By then he had retired to his woodlot without a penny for his contribution to the discovery of the *ob* gene. He says that his pension from Jackson Laboratories will more than take care of his and his family's modest needs. But he worries that the pursuit of cash and fame has distorted the course of scientific inquiry and coarsened its participants.

"Science was once about the free exchange of ideas, about collaboration," he says. "But today it's about intellectual property and confidentiality. It's all about patents, about ownership." Coleman was born to science and he devoted his life to it. But given the squabbling over primacy and lucre, he says he's just as glad to be done with it, just as glad to be out on his lot, chopping wood for free.

Six

THE CLINICAL EXCEPTION

> The obese is . . . in a total delirium. For he is not only large, of a size opposed to normal morphology: he is larger than large. He no longer makes sense in some distinctive opposition, but in his excess, his redundancy.
>
> —Jean Baudrillard

> I'm fat, but I'm thin inside.
>
> —George Orwell

Stephen O'Rahilly is luxuriating in a moment of hard-earned solitude in an Indian restaurant not far from his home in Cambridge, England. Gripping a pint of lager in one hand, a half-smoked stogie in the other, he is dressed for the evening as he was that morning, in baggy corduroys, a shapeless jacket, and a tie of indifferent weave. Sighting me, he snuffs out his cigar, waves over a waiter, and orders more beer and enough curry to feed several maharajahs.

Professor of metabolic medicine at the venerable University of Cambridge, O'Rahilly takes his food seriously. Although he was a tennis champion in his youth, in early middle age he has the look of a man whose relationship with sports is largely confined to spectator status. "One thing I hated about working in America was having people come for dinner, drink one glass of wine, and leave at nine o'clock, saying that they have to work the next day," he says. Clearly, O'Rahilly is made of sterner stuff. No number of late-night pints keeps him from early mornings in his laboratory at Addenbrookes Hospital, a cavernous complex on the city's edge with all the medieval charm of a parking garage. O'Rahilly is busy, almost frantically so since his landmark sightings of human genes linked to obesity rocked the world of metabolic medicine. O'Rahilly was the first to show that Coleman's big idea applied not only in mice, but also in humans.

"I've always considered it distinctly unlikely that there are not genes that affect behavior," he says. "This doesn't mean that every behavior is genetic—for example, that one has a love affair because one is genetically determined to do so. But the idea that a fundamental human behavior like eating is not to some degree genetic, to my mind, is absurd."

Physicians throughout the United Kingdom send O'Rahilly their most baffling metabolic cases, the patients for whom no standard diagnosis will stick. Early this morning his scruffy band of postdoctoral fellows and consulting physicians were puzzling over the mystery of a particularly tragic case: an extended family of Saudi Arabian descent afflicted with obesity so severe that several members had already died of its side effects. The family has no apparent genetic defects, but this seems to galvanize rather than to discourage O'Rahilly, who is used to picking up where others have given up. While the syndromes he studies are rare, their elucidation makes clear that the underlying mechanism driving human obesity is, at its heart, biological.

O'Rahilly is something of a throwback to a time when medical scientists relied as much on observation of patients as on data gathering, testing, and analysis. A humanist in the Rabelaisian mold, he is a passionate intellectual who regards science as an aesthetic rather than as a purely pragmatic exercise.

"It's the beauty part I'm after, the feeling Rosalind Franklin must have had when the structure of DNA revealed itself to her and she said something like, 'it has to be true because it is so beautiful,'" he says. "A Mozart aria is perfect because no one else but Mozart could have done it in the way he did it. And that's what one hopes to do in science, do something in a unique way that is in some sense beautiful."

What O'Rahilly finds distinctly *not* beautiful is society's mistaking overweight as a manifestation of moral turpitude—or, exploiting it as yet another way to bash and marginalize the poor.

"I'm sometimes criticized by so-called liberals, who tell me that I shouldn't be working to validate these nasty people whose disgusting behavior has made them so sick," he scoffs. "People who are not victims of these disorders have claimed the moral high ground, they believe themselves to be virtuous. But the truth is, they're just lucky."

Wallis Simpson, the duchess of Windsor, once famously said, "You can't be too rich or too thin." Anyone who has spent time in regions where one *must* be rich to grow fat recognizes the Marie Antoinette arrogance in this comment. But in the industrialized West, obesity and poverty are to some degree linked, especially in women. There are many sound economic reasons for this, among them that the cost of calorie-dense processed food has steadily decreased, while work and living environments have become increasingly mechanized. Exercise, in a sense, has become more expensive, a luxury that the poor and working classes either cannot access or cannot afford. This in no way implies that overweight is confined to the working class: as the pandemic moves into the twenty-first century, the middle and upper classes are increasingly being pulled into its vortex, but generally speaking, in the West, they are being pulled in at a slightly less rapid rate.

O'Rahilly's impatience with class arrogance runs deep. He grew up in a respectable but shabby Dublin suburb, studying Latin and English literature under the cheerful tutelage of the good Brothers of De LaSalle. His formal scientific education was slight, but by sheer force of will he managed to cram in enough to graduate number one in Ireland in chemistry. "I thought that it was somehow beautiful that chemicals—hormones—could act like messages in the human body," he says. "I had no choice but to become an endocrinologist." At sixteen, an age when many kids are tyranized by hormones, O'Rahilly had determined to master them.

In medical school at the University of Dublin, O'Rahilly drew inspiration from Ivo Drury, a physician who devoted his life to the care and treatment of diabetics. Drury set up diabetes clinics all over Ireland, and spent most of his time administering to patients, many of them diabetic women with complicated pregnancies. "He was a wonderful man, self-sacrificing to a fault," O'Rahilly says of Drury. "His great pain was that he hadn't been able to do more science."

Determined not to suffer a similar fate, O'Rahilly signed on to do diabetes research at Oxford University. His interest was in why and how people develop type 2 diabetes, a condition in which the body loses its ability to regulate blood sugar levels. In 1989, O'Rahilly took a post at Harvard Medical School, in the laboratory of Jeffrey Flier, a research endocrinologist interested in the genetic forces underlying the imbalance in diabetics of the body's normally finely tuned glucose delivery system. This was at the same time Jeff Friedman and Rudy Leibel were tracking the *ob* gene, in the early days of the revolution in molecular biology. The polymerase chain reaction (PCR), a tool now widely used in research laboratories and even in doctors' offices, had just come into use. PCR made it possible to make millions of copies of a single DNA segment in a matter of hours, which made the cloning of genes substantially easier than it had been using older methods. "PCR was ideal for my sort of genetics," O'Rahilly says. "It made it possible for half-trained monkeys like me to do real genetics, because the limiting factor was no longer technical skill, but the quality of the clinical material being studied." The challenge became not so much decoding the genetic material as finding the right genes to decode. At this, O'Rahilly was a master. David Moller, now senior director of the division of metabolic disorders for Merck Research in Rahway, New Jersey, worked with O'Rahilly in Boston on developing PCR-based techniques to examine genes for mutations related to diabetes. "There are any number of (human) genes you could look at," Moller says. "But Steve had a gift for finding just the right patients, and just the right genes."

O'Rahilly was never fully happy at Harvard, or in the cloistered world of American academic medicine. In the United States, he says, practicing medicine is sometimes regarded as a messy distraction from the more fundamental goal of advancing science. This, he believes, is particularly true in molecular genetics, where sheer competitive drive can separate winner from loser in the gold rush to finding and patenting genes. O'Rahilly has always been competitive, but he was not interested in prospecting for genes unless they showed therapeutic promise for diseases he understood and for patients he cared about. He left Boston and returned to England to take a senior fellowship at Cambridge, where he cultivated a tight network of physicians who sent him patients with baffling metabolic syndromes, the "clinical exceptions" O'Rahilly believed would offer clues to the genetic roots of obesity and diabetes.

Among the strangest of these exceptions was a pair of cousins, bright and engaging children whose parents had years earlier immigrated to London from the Punjab, in Pakistan. The father and mother of each child were themselves first cousins. Consanguineous marriage is widely frowned upon in the West, in part because closely related parents are more likely to pass on birth defects. But in some cultures marrying "one's own" is considered preferable to wedding a stranger; such marriages are common in Middle Eastern nations, the Far East, and the Asian subcontinent. Shehla Mohammed, the Oxford University physician and clinical geneticist who referred the cousins, wrote that the cousins seemed to have suffered no ill effects of inbreeding. They would, she continued, be entirely healthy, were they not obese. The older child, an eight-year-old girl, weighed nearly 190 pounds. Despite liposuction and surgery, she could no longer walk and was transported by her parents in a wheelchair. The younger child, a two-year-old boy, weighed a staggering 65 pounds and appeared likely to follow in his older cousin's path.

The parents reported that at the age of four months the cousins became possessed by a voracious hunger, a hunger not unlike that described in Ancel

Keys's starvation experiments half a century earlier. Keys's volunteers had experienced what he called a "semi-starved neurosis," which he discusses at length in his *Biology of Human Starvation.* In a chapter entitled "Behavior in Natural Starvation," he recounts, among the many painful accounts, the experience of Danish Arctic explorer Ejnar Mikkelsen. In 1906, while on a failed mission to find and rescue two other Danish explorers, Mikkelsen's own ship stuck and was then slowly crushed in the ice of North-West Greenland. Mikkelsen and fellow explorer Iver Iverson survived the disaster, and lived together through two brutal winters. Mikkelsen writes that throughout the ordeal he could think, speak, and dream of almost nothing but food. In one scene he describes being forced to kill and eat his last two sled dogs, then falling into a fitful sleep and dreaming of "enormous quantities of food, huge smoking joints, mountains of bread and butter, with great green piles of vegetables and salad. But it is all moving, moving continually; shifting just out of reach." Later, this preoccupation deepens into an hallucinatory obsession:

> For my own part, I can think of nothing but food. At first my thoughts dwell with fond recollections upon all sorts of dishes, but gradually they concentrate themselves upon sandwiches—Danish sandwiches, with no top slice, very different from the dull, dry things one gets in England. Otherwise I have for the last few days dreamed chiefly of enormous juicy steaks as the most desirable human delight, but today it is sandwiches. Why I do not know, but so it is. In particular my fancy turns upon the many packets of delicious food which I have seen given away to beggars, and I grow quite furious at the thought of the contempt with which these gentlemen too often regard such gifts; treasures that I would give years of my life to buy. I remember the neat little packages of sandwiches from my schooldays, and gradually the thought takes possession of me to such a degree that at last I imagine that I am walking in the streets of Copenhagen, eagerly on the lookout for a packet of sandwiches. Sud-

> denly I spy what I am seeking, a little white object lying a little to the right of me. I turn to pick it up before any one else can get it, but, as I stop, my foot strikes against a stone. The shock brings me back to stern reality, reminding me with a painful distinctness that I am in Greenland, far away from Copenhagen and all its sandwiches. . . but the little packages still haunt me.

Reading this and other accounts of near-starvation experiences, one is struck by their similarity—how feelings of shame, love, ambition, and human decency all drop away as the hunger slowly absorbs sensibility and soul. G. B. Leyton, a medical officer captured and interned in a prisoner-of-war camp in World War II, wrote: "None of the other hardships suffered by fighting men observed by me brought about such a rapid or complete degeneration of character as chronic starvation."

Like starving explorers and concentration camp survivors, the Punjabi cousins were obsessed with thoughts of food. But unlike the explorers, theirs was an obsession that food could not quench. They ate more than their siblings, more than their parents, more than anyone could believe. And still they wanted more. Denying food to famished children seemed in every sense cruel, so at first the parents complied with their children's demands. Eventually, though, they took one doctor's advice and padlocked the cupboard door. The children scavenged through the trash for soggy French fries, and gnawed frozen fish sticks from the freezer. There was no stopping them.

O'Rahilly knew immediately that this behavior was beyond gluttony—even greedy children don't eat frozen fish fingers and paw through garbage bins. But exhaustive testing by specialists had found nothing—there was no brain lesion or thyroid tumor, no obvious genetic aberration. O'Rahilly was puzzled and intrigued. This was his sort of challenge, a syndrome so rare and mysterious that the finest clinical minds in England claimed it didn't exist. He agreed to take the case. Just a few weeks before he'd hired Sadaf Farooqi, an Oxford-trained endocrinologist of Punjabi descent. Farooqi had very little laboratory experience, but with her fluent command of Punjabi and her sound clinical training, she was close to perfect for the job.

O'Rahilly had read Jeff Friedman's 1994 *Nature* paper with enormous excitement, and was well aware of what the leptin mutation did to mice. But for three years scientists had searched in vain for a leptin deficit in humans. Leptin deficiency, it seemed, was not a factor in human obesity; in fact, probably quite the opposite was true. Leptin is manufactured in fat cells, so it seemed logical that obese people would have, if anything, an overabundance of the hormone in their blood. And tests showed that many of them did. O'Rahilly didn't really expect Farooqi to find leptin deficiency in the Punjabi cousins. But the leptin blood test was fairly simple to run, and the scientists considered it a sort of warm-up exercise, a first step toward figuring out what was really going on with these children.

"I'd been in the lab only four weeks, and this was the first assay I ever did," Farooqi says. "I had to get familiar with the procedure fairly quickly. We had blood samples from Oxford in the freezer, and I pulled out the older child's first. I ran the assay, and the next morning I met with Steve. His exact words were: 'Oh my God, do you know what this means?'"

The girl's leptin level was not high, as they had expected it would be. Nor was it average. It was nonexistent. O'Rahilly was at first stunned, then circumspect; this had to be a mistake. An earlier examination of the children's genetic material had turned up no sign of a gene mutation. Farooqi, though highly competent, was quite new to the job, and it was probable that she had somehow missed a step. Or perhaps the blood sample was too old, or had been damaged in freezing. Something was amiss.

Farooqi called the children's primary care physician, then drove to London to visit the families. "The parents had almost given up on the medical profession, because they had been told again and again that nothing was wrong with their kids that diet and exercise couldn't cure," she says. "Incredibly, no one had bothered to ask just why it might be that these children ate so much in the first place."

Farooqi drew fresh blood samples from the children and from ten members of their immediate and extended family, packed them in a bucket of ice, and floored her black Toyota MR2 back to Cambridge. She ran the tests that night, and the next morning she met with O'Rahilly. "There was quite a lot of tension," she says. "We all knew that this could be very, very big news."

Once again, the results were remarkable—the children's blood showed not a trace of leptin. O'Rahilly is an excitable man, but an extremely cautious scientist. The absence of leptin did not prove that the children harbored the much ballyhooed obese gene; it merely posed the possibility. There was still work to do. Returning to the lab, he reexamined a gene isolated and cloned from the children's fat tissue, the DNA sequence that normally codes for leptin. There he found what earlier observers had missed: The leptin genes of both children showed the same defect, the absence of a single nucleic acid, guanine, in the part of the gene that normally codes for the hormone. Both children were homozygous for this defect—that is, each had malformations in both copies of their leptin genes. A screening of the parents' blood showed that all four were heterozygous for the obese mutation—that each had one copy of the gene in its normal form and one copy of the gene in its mutated form. The intermarriage of cousins had allowed this exceedingly rare mutation to surface; the parents had passed down two mutated copies of the gene to each of these unfortunate children. Like the *ob* mice, the cousins were constitutionally unable to produce leptin. O'Rahilly and his team had found the first human carriers of the obese gene mutation: these children were fat and getting fatter because their bodies were telling their brains they were starving.

O'Rahilly was aware that leptin injections reversed obesity in *ob* mice, but also that there was no evidence that it would do so in humans. Amgen was at the time running a clinical trial of the drug, but the available results were, at best, mixed; some people injected with leptin appeared to lose weight, but some did not. Even if Amgen's trials had shown leptin 100 percent effective, there was no proof that it would work in these children. The Amgen trial subjects did not harbor the defective gene, and they were not leptin deficient. Leptin had never before been administered to a human who lacked it, because a human who lacked it had never before been identified. Equally important, the Amgen volunteers were adults, mostly overweight men. Injecting children with an untested hormone was a big step, and a potentially dangerous one; children often respond badly to treatments well tolerated in adults. O'Rahilly faced a Hobson's choice; treatment was risky, but without it the cousins were doomed to a life of progressive disability.

"We did experiments in which we gave the children and their siblings test breakfasts, just to see how much they were eating," O'Rahilly says. "We couldn't believe it. The siblings ate normally, but the fat cousins ate huge amounts. The two-year-old would eat 2,500 calories in one sitting—a day's amount for an adult. Later we realized that we hadn't offered him enough. He would have eaten much more were it offered."

Witnessing this made O'Rahilly certain that he had to go forward with the treatment.

Designing an experimental protocol and clearing it through a number of administrative hurdles took an agonizing seven months. In that time, the older cousin celebrated a birthday and gained eighteen pounds. As the pounds piled on, her appetite grew frantic. "She panicked whenever she was out of sight of food," Farooqi says.

Finally the leptin treatment was ready and approved.

They began slowly, with a relatively low dose injected once daily in the nine-year-old. The effect was sudden, and to the girls' parents, miraculous. For the first time in her life, she was willing to push away from the table. She began eating no more than her brothers and sisters, and sometimes less. She stopped begging for food, and slept through the night, rather than haunting the kitchen. "It was one of the most remarkable things I'd ever seen in medicine," Farooqi says.

Gradually, the weight started to drop off. In a year, the girl had lost thirty-five pounds, and was walking without assistance. The following year, after her dose was increased slightly, she dropped another ten pounds. The younger cousin, who began leptin treatments two years later at age four, showed similar improvement.

"This proved that leptin is not a vestigial throwback, sitting around the genome doing nothing," O'Rahilly says. "It meant that leptin not only exists, but that it performs an important function in humans."

More fundamentally, O'Rahilly's team had shown that a tiny defect in a single gene could have a profound impact on human behavior. O'Rahilly's experimental treatment of these young cousins made inescapable that the drive to overeat has deep genetic roots. As Rudy Leibel later put it, "We could no longer get around the presence of genes as a regulator of body weight."

Since O'Rahilly and Farooqi announced their discovery in 1997, about a dozen people worldwide have been found with the leptin mutation. Among these is "Patient 24," who, at five feet six inches tall, weighed 330 pounds and had a fifty-five-inch waist. Described in 1998 by Metin Ozata, a physician in the department of endocrinology and metabolism at the Gulhane School of Medicine in Ankara, Turkey, Patient 24 was a twenty-two-year-old man with no beard, very little body hair, and testicles the size of a young boy's. He showed no signs of puberty. Like the Pakistani cousins, he was born into a highly inbred family, two other members of which—a six-year-old girl and a thirty-four-year-old woman—turned out to also have the leptin defect. The woman had never menstruated. There were other very fat relatives who had died young and childless. Just as are *ob* mice, it seems humans with the *Ob* mutation are infertile. Harvard scientist Rose Frisch's finding of women athletes failing to menstruate now made perfect sense:

extremely lean women simply don't have enough leptin-producing fat on their bodies to turn on their fertility. Humans with the leptin mutation have all the fat, but none of the fertility-inducing leptin.

Jeffrey Flier, O'Rahilly's old colleague at Harvard, had earlier shown that a decline in leptin is a warning signal to the body that something is amiss, and that the response—a heightening of appetite, a reduction of fertility—is protective. The primary role of leptin, then, is not to keep us from getting fat—as Friedman had first surmised—but to keep us from getting too thin, by setting in motion the starvation response so eloquently described by Keys. Without leptin the body does everything possible to minimize energy expenditure, and maximize energy intake. But this doesn't mean that more leptin will reduce eating: the overweight and obese presumably eat more than do normal-weight people, yet they have higher than average leptin levels. Knowing this, it sounds irrational to argue that raising already high blood leptin levels would help anyone eat less.

But some thought that leptin might show a function parallel in obesity to that of insulin in diabetes. Adult-onset diabetics have quite high circulating levels of insulin, yet respond favorably to additional insulin. Raising already high leptin levels in obese volunteers may, it was hoped, perform a similar role and promote weight loss.

Amgen's first test of this hypothesis involved seventy-three overweight volunteers injecting themselves daily with either leptin or a placebo. Only forty-seven people completed the study, in part because leptin injections can provoke skin irritations. Of these, the eight subjects who received the highest (and most irritating) dose lost an average of nearly sixteen pounds, while the twelve on placebo lost about three pounds. But the results were wildly erratic—some people gained as much as twelve pounds on the highest dose, while others lost more than thirty pounds on the same dose. The publication of this finding in the *Journal of the American Medical Association* (JAMA) inspired cautious optimism in some, including Friedman, who was hopeful that leptin would indeed prove analogous to insulin in some subset of patients. Others, like Jeffrey Flier, were less enthusiastic. "For one or two months people considered leptin the Great White Hope," he says. "But we now know that leptin is not very good at preventing people from eating too much."

In late fall 2001, Sadaf Farooqi and O'Rahilly published a follow-up study of thirteen members of three unrelated families of Pakistani origin who were also heterozygous for the *Ob* mutation. They found that people with even one copy of the *ob*gene had lower than normal leptin levels, as well as higher than average amounts of body fat. O'Rahilly concluded from this that a relatively small drop in leptin levels may prompt an increase in leptin-generating fat mass, and that some subgroup of people may benefit from leptin injections.

Still, it is unlikely that leptin per se will prove the fat-buster Amgen and its shareholders had hoped. Obesity is not diabetes—type 2 diabetics gradually lose their ability to make insulin altogether, and with the exception of those with the *Ob* mutation the obese never seem to lose their ability to produce massive amounts of leptin. Leptin injections may reduce eating in some people, but probably not in many and certainly not in the majority. Friedman, who holds a multimillion-dollar stake in the success of the drug, is now looking to see whether leptin might play an important role in *maintaining* weight loss. On its face, this seems like a reasonable assumption. As body fat is lost, leptin levels drop, so by restoring leptin levels, those who have lost weight may have more success sustaining their weight loss. But blood leptin levels vary widely even among people with identical percentages of body fat, and some normal and mildly overweight people have levels that are higher than those of the obese. It is not at all clear that additional leptin on its own will be effective in maintaining weight loss in a significant number of people.

Although leptin is not a cure for obesity, it is critically important: its discovery illuminated countless new avenues for drug development, leads so promising that every major drug company in the world is now heavily invested in obesity research. Jeff Flier is among scores of scientists researching the pathways connecting leptin to fertility and appetite. "It's well within reason that we will know enough about the pathways that affect body weight to control obesity in ten years," he says. "The goal is to find a molecular target in this pathway that you can affect, a drug target that's not negative in some other way." That is, a drug that will curb appetite without doing damage.

More than a dozen critical components in the neurocircuitry underlying appetite control are known, as are leptin-activated signaling pathways among cells. The daisy chain of hormones, receptors, and peptides that regulate body weight is labyrinthine, and involves dozens of genes. Pouring additional leptin into the system may not result in weight loss for most people, but leptin is key. A radical fall in leptin leads to an increase in appetite; it's that simple. Unfortunately, a rise in leptin does not necessarily lead to appetite reduction in humans. This may be because leptin is a necessary but not sufficient component of appetite control; that is, people require leptin to be satisfied, but leptin alone is not enough. Another reason may be that some people—perhaps many people—have difficulty "hearing" the leptin signal.

Cells have on their surfaces molecular gateways called receptors that allow them to admit specific signals from the external environment. When these receptors are missing or are not working properly, important messages go undetected. Friedman and a number of other scientists think that the leptin receptors in obese people may have lost their sensitivity to the leptin signal. How this "deafness" occurred is unclear—it is possible that people were born with the problem, and also possible that they developed leptin resistance over time. Leptin resistance could explain why obese people maintain such high body weights even though they have higher than normal levels of leptin in their bodies.

The leptin receptor gene was a critical piece of the leptin puzzle, and the race to clone it was hotly contested. Friedman's lab was a serious contender, but not, to the researchers' dismay, the winner. First place and the patent went to Louis Tartaglia, head of metabolic medicine at Millennium

Pharmaceuticals Inc., the Cambridge, Massachusetts, company that had earlier lost the *ob* patent to Amgen.

Tartaglia was the very first scientist to be hired at Millennium, a wildly successful biotech firm devoted to finding potent gene "drug targets" for multinational industry partners like drug makers American Home Products, Pfizer, Hoffman-LaRoche, and Bayer. Obesity has been a serious focus at Millennium since its founding in 1993, and the obesity gene that Friedman snared and Amgen bought was one of its most sought-after quarries. Tartaglia's cloning of the first leptin receptor, announced in the journal *Cell* in February 1996, and its subsequent patenting was sweet revenge. Announcing the coup to investors, Millennium attorney Mark Boshar boasted, "This patent further validates Millennium's leadership in obesity research, and represents a clear message from the United States Patent Office that Millennium scientists were the first to discover the role of the leptin receptor in body weight regulation." As Rudy Leibel and others proved, the leptin receptor and the *db* gene product are one and the same. And as Doug Coleman had shown twenty years earlier, *db* mice are fat precisely because they are unable to respond to the weight-regulating product of the *ob* gene.

Not surprisingly, Tartaglia believes that the leptin receptor will prove a more potent target for drug development than will the *Ob* gene itself. So far only three people are known to lack the receptor—members of the same French family who are all massively obese and diabetic. But the ability of the brain to "hear" the leptin signal may also contribute to other, far less rare, obesity syndromes. It is possible that a particular diet—perhaps a high-fat diet—can muffle the leptin signal, thereby increasing our resistance to the molecule, and scientists are working hard to find whether this is indeed the case. What is certain is that an alphabet soup of molecules—neurotransmitters and neuroproteins—is involved in a network controlling the leptin response, and any slips along this pathway can, given certain environmental conditions, fool the brain into demanding more food than the body needs.

O'Rahilly has found evidence of these slips in a number of patients, among these a forty-two-year-old housewife whom I will call Margaret. Margaret came to O'Rahilly's clinic complaining of a lifelong struggle with a hodgepodge of symptoms that doctors attributed to her obesity. Her many dieting efforts had failed, and she harbored little hope that her condition would improve. The mother of healthy quadruplets, Margaret had never menstruated—the babies came only after extensive hormone treatments. She suffered from frequent bouts of incapacitating shakes, sweating, and dizziness. Driving was a death-defying gamble—she got drowsy and once even fell asleep behind the wheel. She has always been obese, but morbidly so only as a child, when she had once spent thirteen months in London hospitals on a supervised starvation regimen. "I was given half a tomato and a lettuce leaf for tea [supper], and was punished for taking a chip from another child's plate," she recalls. "It was hideous."

Chilled by this tale, O'Rahilly asked Margaret to bring him a childhood photo. She returned with a picture of herself at age three being hoisted in her father's arms. The strain on her father's face tells it all—at 84 pounds, young Margaret was, as O'Rahilly puts it, "unbelievably huge." Roughly two and one-half times normal weight, she was unable to walk. As with the Pakistani cousins, O'Rahilly sensed that something was terribly wrong.

But unlike the cousins, Margaret had plenty of leptin. What she lacked was insulin. Insulin is produced in the beta cells of the pancreas, and these cells do not function properly in diabetics. But Margaret was not diabetic. Her blood had loads of proinsulin, an immature and partially effective form of the hormone also produced in the beta cells. Proinsulin is normally broken down into insulin by an enzyme. Probing further, O'Rahilly found that in Margaret the gene known as PC 1, which codes for this enzyme, was defective. Margaret also lacked the enzyme necessary to process POMC, one of two appetite-suppressing proteins in the brain activated by leptin. She had plenty of leptin, but without POMC, her appetite was out of control. O'Rahilly deduced from this that a single genetic defect had disrupted both her blood sugar and her leptin pathway. Publishing these findings in 1997 in *Nature Genetics,* he concluded that "we infer that molecular defects in pro-hormone conversion may represent a generic mechanism for obesity." Margaret's obesity was not the cause of her problems; it was a symptom of a rare genetic syndrome.

O'Rahilly's work has raised concerns—the frightening possibility that what he uncovers about the genetic underpinnings to disease and behavior will be used to control or even cull the less than genetically ideal. This is no idle worry: When the Duke of Edinburgh visited Cambridge for the formal opening of a new clinical research building, he wondered aloud whether it was wise to keep people with genetic disorders alive, as they were likely to breed and perpetuate their conditions.

"Farooqi let him know (albeit subtly) that he was a complete barbarian!" O'Rahilly says. "But the question becomes, is it a good or bad society that gets rid of DNA with an attached disorder? I've been accused of being a biological determinist, and perhaps I am. I think biological determinism has a degree of human decency and kindness about it that is completely absent from the environmentalist view espoused by puritans who want to make everyone behave in a particular way. Obesity is morally neutral, and I have no problem seeing it as a biological problem. I had a girl in my clinic who weighed thirty stone [420 pounds]. The girl's life was a ruin, an utter hell. She never left her house, and she felt like a leper. She'll undoubtedly develop obesity-associated disease, diabetes, and arthritis. Like so many of the people I see, she has a terrible quality of life. Her disorder is a disease as needing of treatment as any other—not sympathy, treatment. Whatever one's ideology, isn't the point to help people get on with their lives?"

Obesity science is a rancorous field in which specialists rarely agree, but O'Rahilly's finding of human carriers of the *Ob* gene, together with the discovery of Margaret's mutation, brought universal acclaim. "This Grail of obesity research has now been found—twice!" raved Rudy Leibel in a *Nature Genetics* editorial. "These observations are important, not because the etiology of most human obesity has now been elucidated—it has not—but because they vindicate an approach to this complex phenotype that emphasizes biology over 'willpower,' and regards body weight as the result of complex interactions between genes and environment, rather than as a psychological aberration of free will."

A schematic of the leptin pathway begins with the fat cell, where leptin is produced, and proceeds to a slew of molecules that activate different groups of nerve cells in the hypothalamic region of the brain. Dozens of brain and gut peptides have been implicated in appetite, and a few have been intensively studied. When a likely candidate is uncovered, it is introduced into the hypothalmus of animals to see whether it has any impact on feeding behavior. The appetite-stimulating neuropeptides are NPY and AGRP, the appetite-inhibiting neuropeptides POMC and CART. Theoretically, leptin both stimulates the appetite inhibitors and inhibits the appetite stimulators. But in reality, these are not separate but equal effects. Eating is too central an activity to be turned off by a single neurochemical switch—there are belts and suspenders galore to make sure that we get enough calories to maintain our body weight. A glitch in a protein that normally turns up appetite will not turn off the eating machine. But a single glitch in a gene coding for a protein that normally turns down appetite—such as Margaret's—will crank up our demand for food.

O'Rahilly has since scrutinized for defects in the DNA of six hundred obese children. Many of these children have symptoms similar to those of the leptin-deficient cousins, but they do not have the *Ob* mutation. What surely many, and perhaps most, of them have is a defect somewhere along the leptin pathway. O'Rahilly determined that twenty-four of these children had a defect in melanocortin-4 (MC-4), and, tracing back to their families, found a total of sixty adults and children with the defect. MC-4 is a receptor in the brain for alpha-MSH, a protein made from POMC that reduces appetite. (A related defect had been identified years earlier in the agouti mouse.) O'Rahilly and others have estimated that as many as 5 percent of obese children carry mutations in their melanocortin system, making it the most common genetic defect yet to be associated with obesity. MC-4 receptors are located on cells in the hypothalamus linked to leptin-responsive neurons, and by boosting their sensitivity, scientists hope to turn down appetite. But again, while increasing appetite seems to be a fairly easy matter, decreasing it is devilishly difficult.

Jeff Flier is among those who believe that melanin-concentrating hormone (MCH)—a hormone made in the lateral hypothalamus that stimulates food intake when leptin concentrations are low—may offer a window into the human appetite system. Eleftheria Maratos-Flier, a Harvard physician and researcher (and Jeff Flier's wife), discovered the role MCH played in appetite, and was the first to study its effect in mice. Maratos-Flier found that mice without an active form of the MCH gene were largely lean, while mice that overproduced the hormone tended to be slightly overweight. This finding fueled a bit of a stir. "Every drug company I know of with an interest in obesity has an MCH-antagonist effort," Maratos-Flier says. If history is any guide, however, it is doubtful that eliminating MCH will have the same impact in humans as it has in rodents.

In the spring of 2001, Jeff Friedman announced in *Science* that his lab at Rockefeller had developed a method of tracing the neural pathways that may control appetite and satiety in rodents. This was exciting news, because it meant that the "cross talk" between various areas of the brain could be traced and perhaps followed to elucidate feedback mechanisms. Friedman is interested in seeing how variations in leptin concentrations can affect higher-order centers in the brain, where decisions to eat—or not—are made. "Maybe signals in the hypothalamus are modulated by signals coming from other regions of the brain," he says. "Maybe in some cases the connections are stronger than others between our conscious selves and our basic drives. Maybe our higher-order cognitive centers send signals to ignore leptin. We don't yet understand this completely, but we will. After all, humans are not metaphysical beings—we are physical beings."

O'Rahilly, too, is optimistic that his work will point toward new treatments for obesity and other endocrine disorders. To that end, he has a relationship with California-based drug maker Incyte Pharmaceuticals. "The serendipitous side effect of my work is that I may be of some assistance in helping pharmaceutical companies develop targets for drugs," he says.

Serendipitous, perhaps, but also felicitous: the announcement of another of O'Rahilly's discoveries related to diabetes sent Incyte's stock price soaring. While infectious disease disproportionately affects people in the underdeveloped world who generally cannot afford drugs, obesity and obesity-linked disorders strike hardest in societies that have at least come to grips with their poverty. For this reason, Wall Street may remain unmoved by a new malaria vaccine, but thrills to every prospect of treatments for obesity and its related disorders. Given the power of the free market to shape scientific research, it is no wonder that since the discovery of leptin, obesity has become one of the hottest fields in science, drawing in the hottest young scientists. Who, after all, can blame them? The worried weighty constitute the largest—and wealthiest—drug market in history. And every drug maker in the developed world wants a share.

SEVEN

COLLATERAL DAMAGE

Being thin is so important that people are easy to exploit, and there is endless money to be made. That's why the diet industry is so dangerous.

—Rose Jonez, former user of the diet drug combination fen-phen

Long Beach, California, is a popular convention town for good reason—the weather is usually obliging, the hotels well equipped, and the cultural distractions few. It is a particularly felicitous venue for the annual meeting of the North American Association for the Study of Obesity (NAASO), a gathering of obesity scientists and physicians, nutritionists, and exercise physiologists from around the world. Conference days begin with scores of participants swimming earnest laps in the hotel pool or trotting the adjacent jogging trail. Later they gather at the continental breakfast buffet, comparing notes over coffee, fruit, and the occasional forbidden Danish served up gratis by one or another corporate sponsor. Not every meal is underwritten, of course, but enough are to give the gathering a slightly sticky feel. NAASO is a respected scientific body, but it is not rich, and by default relies on corporations to pay some of the freight. One cannot help but notice, for example, that a highly regarded obesity specialist is scheduled to testify to the weight loss benefits of soy products at a breakfast sponsored by the maker of said products. This seems particularly curious given that this same specialist had earlier confessed to me his lack of faith in diet aids of this sort.

"We are trying to survive as an organization, but the amount of money available to us is several orders of magnitude less than that of the food and pharmaceutical companies," admits NAASO President Charles Billington, professor of medicine at the Minneapolis VA Medical Center. "This doesn't mean that our ultimate goals are the same, just that we do what it takes to combat the disease." It is no secret in the scientific community that purveyors of weight loss drugs and diet plans feather the nests of the specialists who vouch for them. Nor is it news that corporate patrons expect to get what they pay for: that scientists who find benefit in weight loss products are more likely to enjoy the continued support of the makers of those products. It is in part for this reason that obesity experts tend to interpret their research as hopefully as possible, and to talk up the benefits and downplay the risks of weight loss therapies.

In Long Beach, scientists sponsored by Procter & Gamble Company delivered a paper on the cholesterol-lowering benefits of Olestra, the fat substitute made by Procter & Gamble. Scientists sponsored by the SlimFast Nutrition Research Institute, makers of the meal replacement SlimFast, gave a lecture on the benefits of meal replacements. And Knoll Pharmaceutical Company underwrote what turned out to be the most lavish scientific session of all, on its weight loss preparation, Meridia.

Chairing the Knoll session is University of Pennsylvania psychologist Tom Wadden, an internationally recognized authority on obesity and, not incidentally, a Knoll consultant.

Such arrangements are not rare in academic medicine: many, if not most, high-profile obesity researchers are either consultants to the diet, food, or pharmaceutical industry, or conduct research for those industries. Many do both. The members of the National Task Force on the Prevention and Treatment of Obesity were listed with their affiliations in the September 25, 2000, issue of the journal *Archives of Internal Medi-*

cine. Tom Wadden and nearly everyone else on the list enjoyed one or more industry affiliations.

In 2000, the *Journal of the American Medical Association* published a study of major research institutions, and concluded that more than 7.5 percent of faculty investigators had personal financial ties with the sponsors of their research. Researchers had paid speaking engagements and consulting agreements worth up to $120,000 per year and positions on scientific advisory boards or boards of directors generally paying in the range of $50,000 a year. Fourteen percent of these relationships involved equity ownership.

The problem with this is that researchers supported by industry tend overwhelmingly to hold scientific opinions favorable to that industry. According to a 1996 study in the *Annals of Internal Medicine*, 98 percent of company-sponsored drug studies published between 1980 and 1989 in peer-reviewed journals or in symposia proceedings favored the funding company's drug.

Drug companies make no secret of trading on the good name—and the goodwill—of the academics whose research they support. For example, many of the members of the respected International Obesity Task Force and National Task Force on the Prevention of Obesity also serve on the advisory council of the American Obesity Association (AOA). The main function of the AOA council is to lobby for legislation mandating insurance coverage for weight loss drugs. Its main support comes from drug makers, including Interneuron, American Home Products, Roche Laboratories, Servier, and, of course, Knoll Pharmaceuticals Ltd.

Historically, Knoll's interest in the outcome of the scientific research it sponsors has not been what one could fairly describe as dispassionate. In the late 1980s, the company sponsored a study of its popular synthetic thyroid preparation, Synthroid, by Betty Dong, a clinical pharmacist at the University of California–San Francisco. Knoll assumed that Dong would find its product superior, as at first she did. But her research ultimately determined that Synthroid was no more or less effective than three other versions of the thyroid medication, two of them generic forms that cost half as much. Knoll tried mightily to suppress Dong's findings, and the eventual exposure of this landed the company in an embarrassing class action lawsuit. The lay press expressed alarm at this fiasco, but for insiders to the drug approval system, the only surprise, as one wrote, was why Knoll would be "foolish enough to commission such a 'no-win' study with predictable adverse marketing potential."

In the case of Meridia, Knoll knows it has some serious public relations to do. U.S. sales of the drug soared to more than $108 million in 1998, making it the fourth best-selling drug in the country in its first year of distribution. But sales slipped the following year. Knoll blamed the decline in part on a lack of "consumer awareness" and launched a $50 million advertising campaign directed both at physicians and at the drug's biggest market, affluent middle-aged women. (Direct-to-consumer advertising—or DTC—had its start in the early 1980s, but it didn't really take off until 1997, after the FDA allowed pharmaceutical companies to name both the drug and the disease in the same commercial without reeling off a long list of side effects.) Meridia commercials feature attractive, charmingly dressed overweight women puttering pensively in verdant gardens or strolling the beach hand in hand with supportive-looking men. The ads convey the message that thanks to Meridia these women are satisfied and serene, that they have gained control over their eating habit and of their lives, and that they are loved by the sorts of men who any woman would find attractive. It is all very comforting and reassuring, and it makes sense—the idea of taking control over one's weight problem is so compelling that it is easy to forget that there is a potent medication involved. These ads were, for a time, ubiquitous on television and in glossy magazines, particularly "women's" magazines. But apparently they were not entirely persuasive—the following year, U.S. sales of Meridia dropped once again, to a disappointing $94 million.

Knoll paid handsomely for the opportunity to speak to the NAASO group. Though the session had every appearance of being a purely scientific one, Knoll chose the topics and the speakers. What wasn't revealed that afternoon was that the enthusiasm of Knoll-sponsored scientists for Meridia was not always supported by hard data, even when studies seemed custom-built to showcase the drug. Even in the Knoll-sponsored studies, few people who took Meridia lost more than 10 percent of their body weight over a period of six months, and most lost less than this or none at all. Some gained. When those who did not respond at all to the drug in the first four weeks of testing were eliminated from the study, the average weight loss of "responders" was only about 5 percent more than that achieved by placebo.

Perhaps not coincidentally, 5 percent is the absolute minimum weight loss required of any FDA-approved diet drug.

Obesity Meds and Research News, a subscriber-supported on-line publication, conducted an informal survey of Meridia users in January 1999. It concluded: "Overall, survey respondents did not lose weight successfully taking Meridia." Even those who managed to drop pounds while on the drug found their weight loss taper off and plateau after about six months. Those who stopped taking the drug gained the weight back.

In Long Beach the word is out that Knoll is looking for more sites for its pediatric studies of Meridia, which eventually would involve 450 children at thirty research centers around the country. Several of the conference participants say they want a piece of the action, but voice frustration that children with asthma are excluded for safety reasons. Given that obese children are more likely than other children to have asthma, they fret, excluding asthmatics will make the task of finding study subjects more difficult.

The idea that these children would have to be on multiple drugs for the duration of the study, or that their health might be threatened by the experiment, is not on the table. Nor does anyone voice the concern that most of the youngsters who lose weight while taking the drugs are almost certain to gain it back when they go off them.

Drug makers regard obesity as a chronic condition, like hypertension or diabetes, that requires chronic attention. By this reasoning, weight loss medication must be prescribed for life. The twist is that Meridia and other weight loss drugs are FDA-approved only for limited use by the dangerously overweight, so those who do manage to reach a healthy weight with the help of the drug are not technically eligible to continue taking it. In reality, "successful" losers who graduate from the drug generally regain eligibility when pounds return.

Theoretically, at least, one could go on and off FDA-approved diet drugs indefinitely, gaining and losing weight. Physicians have traditionally frowned upon this "yo-yo dieting" or "weight cycling" as unhealthy, linking it to an increased risk of cardiovascular disease, and a decline in lean body mass. Nonetheless, many obesity scientists today endorse diet drugs with the full understanding that weight loss, if any, is likely to be temporary.

George Bray is among these. Bray is professor of medicine at Louisiana State University Medical Center and Pennington Biomedical Research Center in Baton Rouge, Louisiana, where for years he served as executive director. He is an esteemed, widely published, and much honored physician and scientist, a past president of the International Association for the Study of Obesity, the founding editor of both the *International Journal of Obesity* and of *Obesity Research*, and the author, coauthor, or editor of hundreds of journal articles, abstracts, chapters, and books. He is also a fervent believer in and tireless proselytizer for obesity drugs.

George Bray is seventy years old, tall, slim, and patrician. He is proud to say that he weighs today what he did forty years ago. He finds it frustrating that while at least 36 million Americans have a body mass index of 30 or higher, a measly 2 percent take obesity drugs. Of these, nearly half neglect to get their prescriptions refilled. Bray ascribes this sorry state of affairs in part to a "disappointing performance in the marketing end."

As editor of *Obesity Research,* Bray published articles written by him and by others extolling the virtues of Meridia and other drugs. In its publicity booth in Long Beach, Knoll doled out free copies of Bray's handbook, *Contemporary Diagnosis and Management of Obesity,* in which he recommends that medication be "seriously considered" for all overweight and obese patients over age ten. Bray acknowledges that obesity drugs do not cure, but he considers this a perfectly acceptable condition of treatment. He does not believe that gaining and losing weight repeatedly is a particular danger, or, as others have argued, psychologically damaging. Bray echoed the opinion of many obesity scientists when he wrote in the *Annals of Internal Medicine* that "weight cycling should not impair the use of therapy for obesity any more than the occasional stroke after cessation of hypertensive therapy should block the use of anti-hypertensive drugs."

But while blood pressure medication has shown clear evidence of prolonging life, there is no such evidence for weight loss drugs. Nor is there evidence that Meridia or any weight loss medication is safe over the long haul. The truth is that doctors prescribe Meridia and other weight loss drugs—and the FDA approves them—based on their marginal performance in short-term trials.

Eric Coleman is a medical officer in the FDA's Metabolic and Endocrine Division and the physician largely responsible for FDA oversight of weight loss medications. He seems an eminently sensible man, low-key and thoughtful, not given to sensationalism or overstatement. He tells me that sibutramine, the active ingredient in Meridia, was originally developed as an antidepressant, and was recycled as a weight loss medication when in clinical trials some subjects—while their depression remained—dropped a few pounds. Coleman and other FDA physicians were pretty skittish about the drug, and declined to approve it. Their main concern was that Meridia raised blood pressure in some people. This was troubling, because weight loss is one of the most effective methods of *reducing* blood pressure. Given that high blood pressure is one of the more dire complications of overweight, it seemed unwise to recommend Meridia to obese patients. "I basically had a problem from a scientific standpoint that obese people would get a drug that stimulates the sympathetic nervous system—that's counterintuitive to what you'd want for the overweight," Coleman says.

Knoll did more tests, and resubmitted its application for approval. Coleman was not impressed. In a report summarizing the findings of this second application, he noted with concern that while Meridia did seem to help some patients lose weight, it also raised their blood pressure and their pulse rate. In addition, all the weight was lost in the first six months of the trial, and in some patients slipped back over the following half year of treatment. Meridia did not lower cholesterol levels or improve other blood lipid levels. Coleman concluded that "the data presented in these two preliminary reports appear to lend support to this Reviewer's original recommendation of non-approval."

In 1998 Coleman was overruled by an FDA advisory committee, and Meridia was approved. Coleman was not surprised. "Depending on the prevailing political climate, a drug can be not approved by the medical officer and get approved by the FDA," he says. "This can be okay, because one person shouldn't have all the decision-making capability." Whether or not it is okay in this case, Coleman says, remains to be seen, because Meridia has not been available long enough to reveal its long-term effects. But there are hints of problems to come, for on March 19, 2002, the consumer advocacy group Public Citizen petitioned the federal government to take Meridia off the market, citing nineteen deaths linked to the drug. While the drug had slipped in the U.S. market, it was by then one of Knoll's best sellers internationally, with $118 million in non-U.S. sales the previous year. Bruce Schneider, the former Rockefeller University scientist who, by coincidence, is now among Coleman's physician colleagues at the FDA, says that this makes him extremely nervous. "This drug works in the brain, and my biggest nightmare is that it may contribute to slow memory loss," he says, explaining that gradual loss of memory would not turn up in a short-term study.

Eric Coleman doesn't have nightmares, but he does have concerns. "It's safe to say that if we knew when we approved Meridia what we know now about fen-phen, greater caution would have been applied."

The rise and fall of fen-phen is a cautionary tale—a story that illuminates the pitfalls and limitations of the drug approval process. Fen-phen is a combination of fenfluramine, a sedative approved for short-term treatment of obesity in 1973, and phentermine, a stimulant approved for short-term obesity treatment in 1959. Individually, neither drug is particularly effective for weight loss. But in the late 1970s, Michael Weintraub, then a professor of clinical pharmacology at the University of Rochester, mixed the drugs into a cocktail that seemed to bring out the best in each—the stimulant effect of phentermine toned down the soporific effect of fenfluramine, so people could take the drug without falling asleep on their feet. In 1984 Weintraub published an article in the *Archives of Internal Medicine* promoting fen-phen as a powerful weight loss agent. He followed up with a four-year study of 121 obese patients, supported in part by American Home Products (AHP), maker of fenfluramine. The study was rejected by several journals before being published in 1992 in the *Journal of Clinical Pharmacology and Therapeutics.* The results were not spectacular—on average, women taking the combination drug lost six pounds more than did women taking a placebo. Everyone in the treatment group regained the weight when she went off the medication. And almost nothing was known of long-term side effects. But these caveats were lost in a wave of exuberance. Touted as a weight loss miracle by a wildly enthusiastic press, fen-phen was eagerly sought and eagerly prescribed, and not just for the obese.

"I figured, gee whiz, these drugs have been on the market for ten, twelve years," Weintraub later told *The New York Times.* "Everything must be known about them."

Well, not everything. There had been troubling reports linking fenfluramine, brand name Pondimin, with pulmonary hypertension, a rare but dangerous lung condition. And while FDA-approved drugs—even with unknown synergistic effects—may be legally prescribed "off label" in any combination, neither fenfluramine nor phentermine had been approved for long-term use individually, let alone in combination.

Larry Sasich, a research analyst at Public Citizen, says that many physicians at the time knew of these concerns, but chose to ignore them. "It was a matter of market forces," he says. "They simply needed the money."

Perhaps, but surely just as important was that fen-phen offered hope for discouraged patients, people with compromised health and compromised lives who had struggled for years or decades to lose weight, and couldn't. Honest physicians believed that fen-phen would offer relief to patients they cared about. And honest physicians prescribed fen-phen. But at the same time, the Internet became cluttered with ads from unscrupulous shills hawking the drug to "patients" they had never seen. Strip-mall weight loss "clinics" sold pills to anyone with a checkbook or a credit line. For-profit weight loss centers, like Jenny Craig and Nutri/System, incorporated fen-phen into their programs. *Woman's Day, Reader's Digest,* and other mass circulation magazines and newspapers raved about the drug in ads and articles. And when a chain of California weight loss clinics announced its toll-free number—888-4FEN-FEN—on the Howard Stern show, Wyeth-Ayerst Laboratories, a subsidiary of AHP that distributed the drug, could barely keep up with the demand.

Fenfluramine is made up of two chemically identical molecules in which the atoms arrange themselves in left-handed and right-handed mirror images. Of the two, the right-handed molecule, dexfenfluramine, is the more powerful appetite suppressant and has fewer unpleasant side effects. In the 1970s a French pharmaceutical company developed a technique to produce pure dexfenfluarmine, which it brand-named Redux. Interneuron Pharmaceuticals, a drug distribution company founded by MIT neuroscientist Richard Wurtman, peddled the U.S. license for Redux to Wyeth-Ayerst, a subsidiary of American Home Products. Wurtman, who with MIT already held a patent on fenfluramine, then joined AHP to apply for FDA approval.

In September 1995, the FDA convened an eight-member expert committee to consider the case for Redux. George Bray was one of several obesity experts speaking on the petition's behalf. Also testifying was Joann Manson, then an associate professor at Harvard Medical School, and a consultant to Interneuron. Bray, Manson, and others serving both academic and corporate interests were there to plead the industry case. They were respected scientists who, like many others, had ties to drug companies. That Wyeth-Ayerst and other companies in the business of making diet drugs supported their research did not necessarily compromise their credibility as expert witnesses. But it is safe to say that the drug companies were certain of their expert views, and were eager to see that these views dominated the FDA discourse. At the very least it was well known to Wyeth-Ayerst that both Bray and Manson regarded obesity as a deadly evil.

Other testifiers were leery. They spoke of animal studies that had implicated dexfenfluramine as a neurotoxin that might disrupt the structure and function of certain brain structures. They argued that more studies were needed to tease out the possibility that these effects might adversely affect humans.

Even more concerning were reports from Europe of roughly one hundred cases of Redux-linked primary pulmonary hypertension (PPH). PPH is a rare, incurable condition characterized by breathlessness, debilitating fatigue, and death, usually within three years after diagnosis. Most at risk to acquire PPH are women between the ages of twenty-one and forty, the very population targeted for weight loss medication. The PPH link was so unnerving that the advisory committee turned down the petition to approve Redux for long-term use in the United States.

Given the grave implications of that day's testimony, it would have made sense for the drug makers to return to their laboratories and run more studies of Redux. In particular, they might have wanted to conduct in-depth and carefully controlled tests in humans of more than a year's duration. But neither Wyeth-Ayerst nor Interneuron was willing to further delay the launch of what some analysts estimated would be a $1.8 billion drug. The companies requested a second hearing, and for reasons that were never made fully clear, FDA metabolic disorder chief James Bilstad obliged.

The meeting was scheduled for November, too soon for further tests or even for a serious reconsideration of the evidence at hand. Nonetheless, the testimony was, if anything, *more* damning of Redux than it had been in the first hearing. Paul Ernsberger, associate professor of medicine and pharmacology at Case Western Reserve University, explained that in France, where the drug was approved only for short-term use (three months or less), 20 percent of all cases of pulmonary hypertension, a form of PPH linked to an underlying health condition, had been traced to it. He warned: "If dexfenfluramine is approved by this committee, it is certain that at least a minor epidemic of pulmonary hypertension will likely result." Another testifier pulled out European studies in which subjects who discontinued Redux suffered a weight rebound, ending up fatter than they were prior to taking the drug. Lynn McAfee, a representative from the Council on Size and Weight Discrimination who claimed that her own weight was more than five hundred pounds, said that she and her organization were firmly opposed to the drug. "I believe that in the future drugs will be developed that will be of great use to fat people . . . but dexfenfluramine is not the drug I've been working for. Fat people will have to come up with a considerable amount of money every month of our lives for a drug whose long-term risks have not truly been established and a weight loss that may be barely noticeable to others." Physician Leo Lutwak, an obesity and osteoporosis specialist and the lead FDA medical officer on the case, pointed out that dexfenfluramine seemed to have no positive effect on obesity-linked conditions such as high blood pressure, high blood sugar, or blood cholesterol. He said it was up to "you ladies and gentlemen" to decide whether, given the circumstances, any risk—even a small risk—was worth taking.

Supporters of the petition argued that the risk was very much worth taking. They reminded the FDA committee that 300,000 Americans died each year from obesity-linked disease, and that billions of dollars in health costs were accrued in treatment of obesity-related disorders. They also reminded the committee that few effective alternatives to the drug existed. Under this pressure, the committee's resolve wavered, and then cracked. After a full day of deliberation and soul-searching, a majority grudgingly approved the drug—voting six for approval, five against.

Leo Lutwak was horrified. He assumed that the evidence spoke for itself: Redux showed a high risk for neurotoxicity and pulmonary hypertension. But given the vote, all he could do was insist that Redux carry a bold "black box" border on the prescription label to alert doctors and patients to potential life-threatening side effects. Nervous committee members took solace in the knowledge that a weight loss medication bearing a black box warning would find its way only to the sickest patients—those with life-threatening obesity. But Robert Sherwin, a professor of medicine at Yale who voted no to Redux, later told *The New York Times:* "People had uneasy feelings, I think, on both sides."

In April 1996, Redux became the first diet drug in more than twenty years to gain FDA approval. And thanks to strenuous efforts on the part of American Home Products and its representatives, it went out without the black box warning.

Redux enjoyed the fastest launch of any drug in history. Within months of approval, prescriptions were being written at a rate of 80,000 a week. This seemed only to spur the sales of sister drug Pondimin (fenfluramine), which shot from $48.7 million in sales in 1995 to $150 million in 1996. Over the next twelve months, drug sellers filled eighteen million prescriptions for the two drugs, and sales of fen-phen continued to grow.

Wyeth-Ayerst, distributor of both Redux and Pondimin, left nothing to chance. At the time, it was paying medical publisher Excerpta Medica $20,000 a pop to "ghostwrite" scientific articles on the health risks associated with obesity. These articles were then submitted to medical journals under the "authorship" of internationally known obesity scientists, who were paid $1,000 to $1,500 for the loan of their name. Two of the articles were subsequently published in peer-reviewed publications owned by Excerpta Medical Journals.

One of the review articles, "authored" by Albert Stunkard, extolled the weight loss benefits of Redux. Another, "authored" by F. Xavier Pi-Sunyer, an equally respected obesity specialist at St. Luke's-Roosevelt Hospital Center in New York, argued that losing small amounts of weight is "generally associated with a decrease in risk factors and the alleviation of clinical symptoms," even in cases when some of that small amount is regained. Pi-Sunyer's article was accompanied by an editorial downplaying the risks of Redux.

One can only speculate as to why two highly considered research physicians with lofty academic appointments would agree to sign their names to scholarly articles they did not research or write.

Less than a year later, in July 1997, a Mayo Clinic study turned up twenty-four Redux-linked cases of heart valve damage. (Months later, court documents gave evidence that AHP knew of the heart valve risk long before that time, just as it knew of the PPH risk.). The FDA responded with a public health advisory, advising physicians to closely monitor patients prescribed diet drugs in combination or for longer than three months. By October 1997, when evidence surfaced that as many as 30 percent of users had suffered PPH, valve damage, or other injuries associated with the drugs, the FDA yanked fenfluramine and dexfenfluramine from the market. But for hundreds—perhaps thousands—of people, it was already too late.

Rose Jonez is a former computer consultant and mother of two now living in Aptos, California. Jonez is four feet nine inches tall and has always been on the heavy side, though not morbidly obese. She grew up in Santa Cruz, a beach town where children, like children everywhere, could be cruel. Jonez was not happy. In high school her mother drove her seventy miles north to San Francisco, where she met her first diet doctor. The doctor prescribed amphetamine.

Amphetamine has a chemical structure similar to adrenaline, and a similar effect—it peps you up, sharpens your mental outlook. It was first marketed in the 1930s as Benzedrine in an over-the-counter inhaler to treat nasal congestion, and also by prescription for narcolepsy and for what is today called attention deficit hyperactivity disorder. During World War II, amphetamine kept soldiers alert, and in the 1960s and 1970s it gained popularity among college students, athletes, and long-haul truck drivers. (The Rolling Stones' 1966 hit "Mother's Little Helper" implicated middle-class housewives as abusers as well.) From the early 1960s, it was widely prescribed as a diet aid. But amphetamine is not an ideal weight loss agent. It loses its potency after a few weeks, and to boost the effect the dose has to be continually increased. As the dose rises so does the ante—sleeplessness, anxiety, and, in some cases, psychosis closely resembling paranoid schizophrenia.

Rose was not willing to risk these side effects. She quit taking the pills, and the weight quickly returned. She grew up, got married, had kids, got a great job as a computer consultant. But all the while, she was concerned about her weight. She tried the Beverly Hills diet, the Scarsdale diet, Weight Watchers, Take Off Pounds Sensibly (TOPS), diet shakes, a high-protein diet, a high-carbohydrate diet, a diet that required her to eat mostly pineapple. Some of these regimens worked for a while, but the weight always returned. Finally, a friend mentioned a husband-and-wife team, he a physician, she a psychologist, who supervised a "protein fasting" regimen. It wasn't cheap, but Rose dutifully signed on, paid up, and carefully followed the doctors' advice. She lost lots of weight. But when she returned to her normal routine the weight came back, this time in spades. "It was like a straw had been put in my back to blow me up," she says. The husband-and-wife team advised her to enroll in one of their special programs for the truly determined. More money, more time, more hype, more hope. After two years, Rose says she was feeling better about herself as a human being, but had lost no weight. The dynamic duo—the doctor and the psychologist—suggested fen-phen.

"At first I was pretty reluctant," Rose says. "I knew they were in it for the money. But they were the only people who were interested in helping me lose the weight, so I agreed to take the drug." That was 1992, the year Weintraub published his study, igniting the hope of millions. And Rose's hopes were well placed—she took the pills, dropped thirty pounds and four dress sizes. She bought a new swimsuit and showed it off on a vacation with her husband in the Virgin Islands. She was ecstatic. "It was like a miracle," she says.

But even miracles have their downsides.

In early 1994 Rose started feeling breathless, even on short walks. She figured that maybe she needed to lose a little more weight. So she worked harder to drop the pounds, and continued taking her pills. The pounds didn't come off, and the shortness of breath got worse. That summer, her doctor stopped prescribing the diet pills, citing reports linking a chemical cousin of Pondimin to brain damage in laboratory animals. Rose continued to take the drug on the sly. What the doctors didn't know—and what Rose didn't know—was that she was gaining weight because her body was retaining fluid. "I basically had congestive heart failure," she says. In January 1996, a cardiologist diagnosed her with primary pulmonary hypertension and a damaged heart valve. By then she was too sick to walk, and was spending her days flat on her back or in a wheelchair. "If my kids hadn't forced me to go to Stanford University Medical Center for treatment, I wouldn't have made it," she says.

Rose says she is alive today thanks to Flolan, a drug that drips twenty-four hours a day, seven days a week through a tube implanted directly into her heart. Her life is better than it was a few years ago—she can talk without having to stop to catch her breath, and do light housework. But she will never again hold a job.

"I was a victim of a pie chart," she says. "The drug companies figured, well, some people will get sick or die, there will be lawsuits. But there is so much money to be made that they know in the long run, they'll come out ahead."

It is hard to know whether any drug maker came out ahead on fen-phen. At the very least, the aggressive marketing of the drug presented a thorny public relations challenge and a serious setback to the diet drug industry. In October 1999, American Home Products agreed to a $3.75 billion settlement. At the time, about 4,100 suits involving 8,000 people had been filed against the company, making it one of the largest product liability cases in history. About $2.8 billion went to settle claims by people who, like Jonez, had already been injured. The company also set up a $1.2 billion fund to medically monitor users of the drug who did not yet show signs of injury, and set aside an additional $410 million to cover plaintiffs' attorneys' legal fees. By early 2001, estimates of the final amount needed to settle all litigation, including the nationwide class action settlement, individual settlements, and the cost of surgeries and legal fees, had soared to more than $12 billion, a staggering sum that would bankrupt many businesses. But Wall Street didn't blink. Soon after AHP announced that it would be setting aside the $12 billion, its stock started to climb, and the company announced a dazzling 17 percent leap in operating profits for the previous quarter.

The pharmaceutical industry is by far the most profitable in the United States, with profits exceeding even those of commercial banking. The marketing budgets of the major drug companies outstrip their research and development costs, and an increasing chunk of that marketing is directed not at doctors, but at consumers. The purpose is to get the patient to ask for the drug, rather than to wait for doctors to suggest it to the patient.

Doctors commonly know little more about a new drug than what their patients and the drug company sales force have told them. This means that the benefits of a drug are trumpeted far louder than the risks. And for drug companies, new obesity drugs are a gamble well worth taking: with an estimated 1.1 billion people overweight or obese worldwide, the potential profits to be made on a successful drug are beyond imagination.

Xenical, the newest FDA-approved diet aid, is, at this writing, the best-selling prescription weight loss medication on the market. Approved by the FDA in April 1999, it had U.S. sales that year of $146 million, translating into 1.5 million dispensed prescriptions. The following year U.S. sales soared to more than $207 million. (It was already selling briskly in Europe, where the European Commission had approved it a year earlier.) Today, Xenical is Roche Pharmaceuticals' third largest-selling drug. Part of the reason for its continued success is that doctors are not afraid to prescribe it.

Meridia, fen-phen, and most other diet drugs act in the central nervous system to reduce appetite, the thought of which makes many people uneasy. Xenical acts in the gut, and that makes people less uneasy. Its active ingredient, orlistat, inhibits the enzymes in the gastrointestinal tract that break down dietary fat into absorbable components. Popping Xenical before a meal causes about one-third of consumed fat to flush through the gut undigested. This may result in mess and embarrassment, but it doesn't involve mucking around in the brain. Doctors feel better about prescribing Xenical, and patients feel better about taking it.

Xenical seems to embody our most indulgent weight loss fantasy—a pill that stops the body from absorbing calories from food we have had the pleasure of eating. But Xenical is not really about having one's cake and eating it too, because if you swallow Xenical and the cake you'll regret it—the drug provides a powerful incentive to limit one's consumption of fatty foods. (One obesity researcher likens Xenical to "gastric Antabuse," the aversion drug that discourages alcohol abuse by altering one's ability to metabolize it. People on Antabuse who drink experience a litany of unhappy side effects, ranging from flushing, throbbing in the head and neck, respiratory difficulty, nausea, vomiting, sweating, thirst, chest pain, blurred vision, and mental confusion.) Roche is up-front about the less desirable effects of Xenical. In advertisements, it states flatly: "Since Xenical blocks about one-third of the fat in the food you eat, you may experience gas with oily discharge, increased bowel movements, an urgent need to have them and an inability to control them, particularly after meals containing more fat than recommended." Suffice it to say that orlistat chased by a cheeseburger and fries can result in an unpleasant evening.

Xenical packs a one-two punch—it prevents the body from absorbing part of consumed fat and it discourages those who use it from eating more than the recommended amount of fat in the first place. What part of this equation helps people drop excess pounds is unclear, but it does seem that using the drug religiously—before every meal—helps some people lose as much as 10 percent of their body weight. For most people, though, the loss is much less dramatic, averaging about six and one-half pounds more weight loss than experienced with a placebo. Most of the pounds drop off within the first six months. Maintaining this weight loss costs roughly six dollars a day, or more than two thousand dollars annually. Most, if not all, of the weight returns when the drug is discontinued. Those who stick with the pills are often frustrated to find the pounds slipping back after about two years, when Xenical starts to lose its effect. Meanwhile, reports of serious side effects have trickled in, including adverse complications for diabetics and liver failure.

No diet drug is FDA-approved for indefinite use, but this is no real barrier, because consumers need not bother with doctors or prescriptions to score their fix of diet remedies: Xenical and Meridia can be purchased with a credit card on the Internet. And by far the most popular weight loss potion, ephedra, does not require a prescription or a doctor's okay.

Ephedra is an herbal supplement sold on-line as well as off-the-shelf in drug and health food stores. A stubby, shrublike plant, ephedra has been a mainstay in Chinese herbal remedies for more than four thousand years. It is also harvested in the southwestern United States, and, because it grows in Utah, is sometimes called "Mormon tea." Classified as a food rather than as a drug, ephedra is not regulated by the FDA or by any other federal agency, and is the active ingredient in more than two hundred products, most of them promising weight loss or, to a lesser degree, an energy boost.

In December 2000, the *New England Journal of Medicine* published a review of ephedra-linked health problems. Were it of anything other than a diet aid, the review would have been front-page news. Of 140 adverse events experienced by ephedra users, 31 events were deemed definitely and 31 were deemed possibly linked to the drug. These included stroke, seizures, hypertension, and tachycardia. There were thirteen cases of permanent disability and ten deaths. The authors concluded that "our findings arouse concern about the risks of these products, given that they have no scientifically established benefits." Ten months later, Public Citizen filed a petition with the FDA demanding a ban on ephedra, citing as support reports of 1,398 adverse effects, including eighty-one deaths, between January 1993 and February 2001. Still, proving that ephedra *caused* these heart attacks, seizures, strokes, and deaths was close to impossible—who was to say that the poor souls wouldn't have died even more quickly without the drugs?

Ephedra's active ingredient is ephedrine. Synthetic ephedrine is a staple in prescription asthma medications and on "crash carts," the emergency room trolleys filled with drugs and devices used to shock the near-dead back into this world. Like codeine and cocaine, morphine and quinine, ephedrine is an alkaloid, and its stimulating effect on the central nervous system—raising cardiac output, blood pressure, and heart rate—is similar to that of amphetamines. It is sometimes called "the poor man's speed."

Ephedra speeds the resting metabolic rate—the rate at which humans burn the calories necessary to carry on the basic functions of living. The idea of burning off extra calories without doing extra work has great appeal, and drug companies have for decades hungered after a weight loss elixir that would uncouple the burning of calories from the performance of work. But ephedra does not uncouple these functions, and it is the very side effects of ephedra—the nervous jitters, pounding heart, thumping blood pressure—that speed metabolism. In the most popular ephedra preparations, this effect is enhanced by the addition of caffeine.

Metabolife 356 is such a product. Metabolife has gone further faster than any other company in the weight loss industry, boasting a billion dollars in sales in 1999. Company founder Michael J. Ellis describes on his website how in 1989 he lovingly brewed a potion of vitamins, minerals, and herbs to help his father regain energy and enjoy life while waging a battle against bone cancer. His father lost the fight, and in 1990 Ellis, a former California police officer, chauffeur, and real estate agent, was implicated in a methamphetamine lab bust. (Methamphetamine, better known as meth, is an illegal street drug chemically related to ephedrine.) In 1992, while still on probation, Ellis and some colleagues marketed Fosslip, a bodybuilding formula containing, among other things, an ephedrine derivative. Fosslip flopped, so in 1995 Ellis cut off his former partners (who later sued him) and began peddling Fosslip as Metabolife. Again, sales were lackluster, but Ellis had a brainstorm—if he couldn't improve his product, he could improve his marketing plan. He made customers independent dealers, and built a legal pyramid scheme. There are today about 50,000 dealers hawking Metabolife in the United States, through mall kiosks, the Internet, and hand-scrawled signs tacked on telephone poles along the roadside. Ellis, who lives in San Diego, has become a rich and very powerful man. He donates generously to political campaigns, and threatens suit against journalists and doctors who challenge the efficacy and safety of his products.

On August 8 and 9, 2000, the Department of Health and Human Services held a public meeting on the pros and cons of ephedra-based weight loss products. George Bray was in attendance, this time speaking on behalf of Metabolife. Bray has spoken and continues to speak publicly on the promise of ephedra and caffeine combinations for weight loss. At the hearing he was not moved by the testimony of other experts of the less-than-thrilling record of this cocktail as a weight loss aid. He was not moved by other experts who raised concerns over the association of ephedrine with heart failure, seizures, stroke, and even psychosis. He was not moved when he heard that in the previous nine years the FDA had received reports of 1,176 adverse events associated with ephedra and, in particular, ephedra and caffeine combinations, and that some of these complaints were of death. He was not swayed by the professor of neurology and pharmacology from Johns Hopkins Medical School who pointed out that "deconditioned overweight individuals would be expected to be more susceptible to the cardio- and cerebral vascular complications" associated with ephedra. Throughout, he maintained his enthusiasm for Metabolife just as he had earlier for fen-phen. "From the data I reviewed," he said, "I must conclude that over-the-counter preparations of ephedrine and caffeine are safe when used according to the directions."

Bray is an astute and thoughtful man who has put together what, to my eye, appeared to be the world's most complete private collection of books on the history and science of obesity. He has devoted his career and much of his life to thinking about, researching, and treating a highly stigmatized condition, a condition that many physicians of his generation chose to ignore or to ridicule. He has lived long enough to see dozens of theories of obesity come and go, and he is understandably frustrated. Like many physicians, he considers obesity a disease. Like all physicians educated in the American medical school system, he was trained to cure with drugs.

* * *

Bruce Schneider has some thoughts on all this. For over a decade, Schneider has worked as a medical officer in the endocrine metabolic division of the FDA in Rockville, Maryland, where part of his job is overseeing review of diet drugs. "The public," he says, "has no idea what demons I have to work with." Schneider says that the drug companies and "the people they've paid off in academic medicine" have portrayed "obesity as a lethal threat, like anthrax. Anything that 'works' should get on the market, because the alternative is that thousands will die of the side effects of obesity. But none of these drugs improve health over the long term. And the risks of these drugs are very real. The risks are what keep me up at night."

All drugs carry risks, and obesity drugs should not be held to a higher standard than are other drugs. But the drug industry is well aware that balancing the risk of weight loss medications against life-threatening obesity is disingenuous, because people with life-threatening obesity are by far not the only ones who take the drugs. As the drug industry well knows, many people who take diet drugs are not at a high enough level of weight-linked health risk to be treated with any weight loss drug.

Weight loss drugs are an iffy proposition for everyone. Unlike leptin, which resolves a rare genetic defect, obesity drugs are not meant to treat a specific pathology. Rather they are designed to interfere with what is essentially a healthy, smoothly running system. The labyrinth of genes, peptides, and hormones regulating food intake is dense and byzantine, extremely difficult to fool or to manipulate. Knocking out one or a couple of components of this system with drugs is unlikely to work, because eventually other components step in to take their place. The most powerful appetite stimulant known is neuropeptide Y (NPY), a molecule that, when injected into the brains of rodents, results in a feeding frenzy. Leptin interferes with NPY production, and for a while scientists thought this was how leptin controlled appetite. There was hope that a drug that interfered with NPY production would reduce appetite, and perhaps even cure obesity. In May 1997, scientists at the University of Washington announced that they had engineered mice without the gene to make NPY. This was exciting news. The disappointing news was that even without any NPY at all, these altered mice ate as heartily as normal mice.

Mother Nature wants her children to eat. Knock out a satiety gene like *ob* and an animal will eat insatiably. But knock out an appetite gene, and animals continue to eat normally. For this reason, appetite-controlling drugs like fen-phen and Meridia, while they may work for some for a year or two, are not likely to permanently alter the eating habits of millions. The brain mechanisms that control food intake generally work perfectly well in most people: the rate at which we use energy varies precisely with our lean body mass. Generally speaking, the more lean body mass, the higher the metabolic rate. It's that simple. The ironic thing—the thing that most people don't consider—is that in order to support their much higher than average total body weight, the obese tend to have, in addition to more fat, a *higher* than average *lean* body mass and a somewhat *higher* than average metabolic rate. On reflection, this makes perfect sense. Consider the extreme example of Jon Minnoch, the Bainbridge Islander who at his death in 1983 weighed a staggering 1,400 pounds. If, as some speculate, Minnoch was 80 percent fat, this leaves 320 pounds for bone, organ, and lean muscle mass. The energy expended to support that muscle mass had to be enormous.

For decades, scientists have searched for a way to safely speed up human metabolism by tinkering with the chemicals in the brain that control our rate of energy burn. One of the favorite targets has been the beta-3 receptor that mediates the activity of brown fat. The sort of fat we think of as fat—white fat—stores energy. Brown fat burns energy. Brown fat is plentiful in rodents, and in human newborns, where it seems to function only in the early weeks of life, until the infant develops other biological mechanisms for staying warm. In human adults, brown fat is widely dispersed throughout the body, and does not seem to play much of a role. Inside brown fat is a protein, the uncoupling protein (UCP) that allows the cell to dissipate energy by disarming the mechanism that generally turns energy into fat. Uncoupling proteins turn a fat cell into a tiny furnace that burns rather than stores heat. In the past decade, scientists have discovered three new varieties of UCP—UCP2, UCP3, and UCP4—in human adult cells, and there is hope among some that there may be a way to stimulate these proteins with drugs to cause fat cells to burn, rather than store, energy. The benefit of such a system is that it could avoid the brain entirely and work directly on the fat cells, which would make a lot of doctors—and patients—breathe easier. The dream, of course, is to create a drug that will allow humans to eat anything and everything that they want, but simply burn it off as heat.

There are several problems with this, the first one being that mice without UCP do not get fat, which suggests that increasing UCP activity in humans may not have the desired effect. Even if it were possible to speed up metabolism by increasing UCP activity, the side effects are likely to be daunting. All UCP-generated heat has to go somewhere, and it is likely that anyone taking such a drug would experience a slight—but unpleasant—increase in body temperature, akin to a never-ending hot flash. There would also be an increase in the formation of free radicals, the nasty chemical rogues thought to hasten aging.

The obese remain obese despite their more active than average metabolisms, and so ratcheting metabolism up still further seems unlikely to have a salutary effect. What is much more likely is that a chemically induced increase in metabolism will result in compensatory eating—that the consumption of drugs will increase the consumption of food. Compensatory eating may be a good thing for the economy, boosting, as it would, both the drug industry and the food industry. But given the growing body of research linking longevity with low calorie consumption, perhaps it is worth considering whether it would be a good thing for the human race.

Obesity has many causes and no easy remedy. Often, tinkering with one aspect of human physiology, with brain chemistry, or metabolic rate, produces unanticipated consequences that themselves need fixing. This has led to a call for combination drug treatments: a combination that may involve one medication to raise metabolism, another to dull appetite, perhaps a third to increase the burning of dietary fat and inhibit the production of body fat. These combination drugs will, of course, have costs. And they will, inevitably, have dangers, which will also require treatment. So there will be more drugs to treat the side effects of these drugs, and so on.

Combination drug therapy may work for some, but it doesn't seem like a particularly hopeful concept, or a genuinely realistic solution. The surge in the past decade of diet drugs has been accompanied by the greatest surge of obesity in history. Genetic studies give evidence that billions of us are prone to becoming overweight. Rather than crashing ahead toward the next drug "breakthrough" to rescue us from this perilous course, perhaps it is time to reconsider. Perhaps it is time to retrace our steps and look back at what brought us here in the first place.

Eight

SPAMMED

JOHNSON: He eats too much, Sir.
BOSWELL: I don't know, Sir; you will see one man fat, who eats moderately, and another lean, who eats a great deal.
JOHNSON: Nay, Sir, whatever may be the quantity that a man eats, it is plain that if he is too fat, he has eaten more than he should have done. One man may have a digestion that consumes food better than common; but it is certain that solidity is increased by putting something to it.

—from Boswell's *The Life of Samuel Johnson*

Approaching Kosrae, Air Micronesia Flight 957 banks gently over twin volcanic peaks cloaked in rain forest lush with wildflowers and framed by a crystalline sea. Touching down on the skinny coral landing strip, we deplane into thick jungle heat and the unctuous scent of frangipani. In Kosrae, a tiny island state of fewer than eight thousand souls, new arrivals are a thrilling curiosity, and it seems that half the island has come out to greet us. Steve Auerbach, an American epidemiologist, and I press through the thicket of Hawaiian shirts and swaying muumuus toward Bill, the man with the keys to the rental car, an American import with plenty of mileage and one bald tire. We shake hands and sign forms without reading them. Bill grins and hands Steve the keys. Groggy from the 14,000-odd-mile journey, we stagger into the compact and grope our way down the coastal road. Starving mutts too lethargic to bark line the roadside like pilgrims. Trees bend nearly double under loads of papaya and breadfruit. The sea, dazzling and empty, beckons, but we do not let it distract us. We are on our way to a funeral feast.

We arrive to find the wake in full swing. Men huddle in lawn chairs playing cards while toddlers squat, transfixed, around a screen blaring videotaped loops of cartoons. Hovering women fill plates and wipe toddlers' faces. Men huddle and gossip. There are perhaps a hundred people here, and even the dead man's wife looks bored. The deceased, buried four long weeks ago in a nearby crypt, seems almost beside the point.

Kosraeans die young—the man in the crypt was fifty-six—but not from the developing world's usual suspects. There is no famine here and, with the notable exception of upper respiratory infections, there is little evidence of the relentless microbial scourges that cut life short in, for example, sub-Saharan Africa. The big killer of adults in Kosrae is what epidemiologists call the "New World Syndrome," a constellation of maladies brought on not by microbes or parasites, but by the brutish assault of rapid Westernization on an unassuming traditional culture. Diabetes, heart disease, high blood pressure—scourges of affluence that long ago eclipsed infectious disease as killers in the developed world—have only recently turned their attention here.

We sit with the dead man's brother-in-law, who tells us that he too expects to die soon. He was once fat like the others, he says, but the diabetes has made him thin. His sister's husband died of heart disease, and he himself will go of diabetes. "But I am fifty-seven, an old man, so this is of no matter," he says. Nodding sagely toward the card-playing young fathers clustered nearby, he says that on this island it is not uncommon for a man to fall prey to a heart attack before his thirtieth birthday. A small girl toddles over to offer us a paper plate of bananas. The bananas, thick yellow fingers, are the sweetest I've ever tasted. Dozens of bunches of these and other banana varieties hang like banners around the place, gifts from mourners who have come to pay their respects. Almost no one seems to eat them. They are beginning to rot in the sun.

Kosrae was once a mighty kingdom, with the city of Lelu its capital. Lelu is still the state's largest and most densely populated village, a jumble of tin-roofed concrete huts connected to Kosrae proper by a causeway. I drive there later that week, to visit the ruins of ancient Lelu. The old city was built sometime between A.D. 1000 and 1400 of immense pentagonal basalt "logs" shuttled by canoe from halfway around the island, and assembled into one hundred or so individual compounds, some with walls reaching as high as twenty feet. It took strong men to raise these walls, strong beyond anything I can imagine. A labyrinthine stone path leads past canals and burial tombs tangled in jungle foliage, past the ruins of once-stunning ceremonial compounds. The German artist Friedrich von Kittlitz visited Lelu in the mid-nineteenth century and marveled at its scalloped roofs and graceful gables "painted red like canoes, with pleasing white decorations," its furnishings of the "highest simplicity." These looming remains are silent with the secrets of a once-great civilization abandoned and largely forgotten.

I contemplate this while sitting on a boulder cushioned with moss. The place is impressive—and humbling. It is also unbearably humid, and a steady stream of salty sweat obscures my vision and sours the mood. I cut short my reverie and duck into a nearby general store for a cold drink. Inside my eyes adjust slowly to the gloom. There are no bananas here, no breadfruit, no papayas, no coconuts. What I see is row after row of canned goods: Spam and corned beef and Vienna sausages spiffy in couture tins. There are cake and muffin mixes from the United States; ramen noodle soup from the Philippines; fifty-pound sacks of polished white rice; flats of soda and Budweiser beer; and shelf upon shelf of candy bars, potato chips, and pork rinds. In a far corner an entire freezer is reserved for turkey tails, a fatty, gristly hunk of the bird not widely popular in the United States. The turkey tail freezer is empty. The delicacy is so popular, I am told, that the month's shipment ran out weeks earlier. Another boat is due next week. I should come back then. I had hoped for coconut milk, but I settle for a Coke.

Robert Louis Stevenson called the coconut tree the "giraffe of vegetables," a freakish, almost ludicrous plant. It was at one time to Micronesians what the buffalo was to Native Americans, which is to say, almost everything. It offered shade from the sun and thatch for housing, clothing, and baskets. The nuts were both food and drink. But in Kosrae today coconuts are mostly a liability, leaving their mark on the rumpled hoods and smashed windshields of what seems like every tenth car and pickup truck. Thick rings of coconuts circle the coconut palms outside my hotel, rotting in the sun, like the bananas.

No one bothers to gather them.

I visit every grocery store on the island, and find no coconuts. But there are plenty of salty, sweet, and fatty imports, pudding mix, macaroni and cheese, and Day-Glo breakfast cereals. There is canned tuna and—incredibly—something called "artificial coconut flavoring." There is no fresh fish or breadfruit, mango or papaya. This is not to say that such things aren't sold on the island. A smattering of fruit stands hawk sacks of Kosrae's famously luscious green tangerines, juicy in their baggy, elephantine skins. There is a fish shack or two. But the fish stands are usually empty or closed, and the fruit stands forlorn. I am told that Kosraeans don't often patronize these places. Not so long ago people here grew their own produce on family plots, and pulled tuna and reef fish from the sea. But this is now the exception, at best a weekend diversion. Modern Kosraeans don't have the time or the energy to farm or to fish—they are busy with their office jobs.

Kosrae was once a fiercely independent kingdom. Today it is one of four island states that make up the Federated States of Micronesia (FSM), the largest and most populous political entity to emerge from the Strategic Trust Territory of the Pacific Islands, the agreement that placed the islands under United States administration after World War II. In 1986 Micronesia signed a compact of free association with the United States, which dissolved its trust status. As part of the agreement, the United States remains FSM's chief benefactor, supplying the bulk of its revenue, about $100 million in aid each year, in exchange for free access for the U.S. military. The bureaucracy required to manage and distribute this windfall has grown into Kosrae's single largest employer. Few jobs here demand the level of skill or physical effort required by traditional fishing and farming. And people tend to move slowly, as befits the tropical climate. Physical exertion has been further discouraged by the steady importation of cars and trucks, some purchased with the help of government money, and by the expansion of the island's encompassing coastal road. The coastal road is pitted with potholes, and drivers swerve wildly to skirt craters the size of large pigs. An American soldier stationed here warns me away from jogging—he and many of his fellows have twisted their ankles in those craters. I try morning walks, but quickly discover that this will not do. To walk in Kosrae is to announce that one is too poor to own a car, and Kosraeans, renowned for their generosity and kindness, offer lifts to every casual stroller.

This newfound convenience has come at a high price in Kosrae, as a tour of the state hospital reveals. A low-slung concrete structure with greasy windows and no air-conditioning, the hospital has a view of the sea, but no library. It is poorly equipped to handle anything but basic health needs. Serious cases are airlifted to Guam or to the Philippines. Though he assured me that his staff was excellent and his facilities adequate, the hospital director, a former vice president of the FSM, takes no risks with his own health. His wife told me that both she and her husband travel abroad even for routine checkups.

The hospital inpatient ward has perhaps two dozen beds, of which nineteen are occupied when I visit. Thirteen people are here for complications of diet-related diabetes, and two for heart conditions. Dr. Paul Skilling, a Kosraean physician, laments that diabetes, hypertension, and heart disease are as common as coconuts on this island and even more dangerous. Another physician jokes that even health care professionals succumb. "Look at me," he says, pointing to his paunch. "I am myself obese. My body mass index [BMI] is 32. How long before I have these diseases?"

The good doctor is indeed obese, but his BMI is only slightly higher than average for a Kosraean man of his age. In 1993–94 the FSM department of health, with funding from the Centers for Disease Control and Prevention

(CDC), screened all the adults on the island and found that nearly 85 percent aged forty-five to sixty-four were obese. It is perhaps not surprising that many of these people were also diabetic, and that more than a third suffered from high blood pressure. Dr. Vita Skilling, the island's chief of public health and community health (and no relation to Paul), says that efforts to reverse this trend have been disappointing.

"Here you buy imported food in the store to show that you have money," she says. I later hear that some Kosraeans scatter empty Spam and corned beef tins around their property as status symbols. But whatever panache imported foods carry is not really due to their extravagance—in fact, quite the opposite. Imported food is cheap, and children in Kosrae have learned to prefer and demand it. A thirty-pound case of turkey tails imported from the United States costs eleven or twelve dollars, a locally grown chicken five dollars. A fifty-pound bag of polished white rice can feed an extended family for weeks, and it doesn't rot in the tropical sun. Rice is not native to the island and it is not grown here, but when there was a rice shortage a few months earlier, some children actually went hungry rather than eat breadfruit or taro.

"We eat white rice for breakfast, lunch, and dinner," Vita Skilling says. "We make the excuse that children prefer it, but the real reason is that it is so easy to cook. It's so much easier to cook rice for the day than to go to the farm to pick fruit. The imported food we eat is too high in salt and fat and we eat far too much of it. And we've gotten used to sugar—some older people can no longer drink plain, unsweetened water."

Dr. Skilling is not too proud to make mention of what to my eyes is obvious—that even she, with all her knowledge, is fat. She has no immediate plans to remedy this, but she does worry. "Diet-linked diabetes is the number one health problem in Kosrae," she says. "It cuts us down."

Uncontrolled diabetes leads to circulatory problems, foot ulcers, skin abscesses, and eventually to heart disease, kidney failure, and blindness. It is a First World disorder that increasingly has come to plague the developing world. In Kosrae, 90 percent of hospital surgical admissions are diabetes-related, often amputations necessitated by vascular breakdown. There are more cases of renal failure than the hospital can handle, and cardiovascular disease is pervasive.

"People deny having symptoms until they are very sick, and then they demand a miracle to fix it," Vita Skilling says. "And they are fatalistic; they expect to get diabetes because so many of their friends and neighbors and family have gotten it." Many here choose not to have their diabetes treated. At age fifty-five or so they begin to feel that they are living on borrowed time. They are not greedy for longevity.

I stop at a fruit stand for a bag of tangerines, and notice a white man in surgical scrubs hovering over a pile of papayas. He tells me that his name is Ken Miklos, and that he first came here from his native California in 1995 on a mission with the Seventh-Day Adventists. He liked the place—in particular the surfing, which is wild and lonely; Kosraeans respect and work the sea, but rarely play in it. When his mission was over he returned to California to complete dental training, but he couldn't get Kosrae out of his mind. So after a couple of years he came back to set up practice at the state hospital and to marry a Kosraean woman. He has been surfing and pulling rotten teeth ever since.

"I see kids with twenty primary teeth, every one of them not just with cavities, but bombed," he says. "We're talking abscesses, six to eight at a time. I have to yank 'em. In church they have candy to keep them quiet. They're nursed with sugar water. They have calcium deficiency and vitamin A deficiency. Half the teenage girls are obese. They eat U.S. imports loaded with fat. It's tough to change your lifestyle. They are sick, their legs are amputated, they're dying. But it's what they're used to."

Obesity in Kosrae seems ubiquitous, but it pales in comparison with the problem in Nauru, a tiny island republic a few hundred miles to the southeast. For millions of years, billions of birds nested on this atoll, depositing tons of guano that leached through the coral to form a hard, colorless rock that is 85 percent phosphate. A generation ago, Nauru's citizenry grew rich from the mining of this phosphate, and Nauru was dubbed (by those few who had heard of it) the "Kuwait of the South Pacific." Soon mining eclipsed fishing as the major revenue source, and the island's few patches of arable land were laid barren by the pounding weight of mining equipment. With farming and fishing all but forgotten, Nauruans today subsist almost entirely on imported food, soft drinks, and beer. Prosperity brought Nauruans Japanese televisions, German luxury sedans, and Australian filet mignon. It also brought the worst of American cuisine—processed foods with plenty of fat, salt, and sugar. Nauruans have one of the highest rates of diabetes and gout on the planet, and a life expectancy of fifty-five. They are known for the ferocity of their dogs, and for being suspicious of strangers. On Kosrae I met a German museum curator, an expert on South Pacific culture, who had stopped in Nauru for a professional look-around. He told me that he had planned to spend a few days on the island, but that he fled the morning after he arrived. He was afraid of the dogs, he said, but he was even more afraid of the men. "They were so fierce looking," he told me. "And so enormous."

Fifty years ago the people of Nauru might have easily been mistaken for the mostly thin people of Kiribati, a republic of thirty-three small islands scattered like stars three hundred miles to the east. Robert Louis Stevenson lived there and described "days of blinding sun and bracing wind, nights of heavenly brightness." Kiribati is still stunningly beautiful, but it is also among the poorest nations on earth. There are few cars on the archipelago, which is just as well, as only two of the islands can boast even short stretches of passable road. The people eat mostly fish, papaya, pumpkin, and breadfruit with their rice. They are known for their friendliness and hospitality, and cling tightly to their traditional culture. They have one of the lowest rates of noncommunicable disease in the South Pacific. In this they resemble the neighboring Nauruans not a whit.

Scientists have studied the health status of Pacific Islanders for decades, and have noted the explosion of diet-related disease. But a CDC-supported effort in 1994 was the first to offer systematic health screening of adults in the Federated States of Micronesia. Steve Auerbach was in charge of that screening. A Manhattan-based medical epidemiologist and officer in the U.S. Public Health Service who worked in the FSM from 1991 to 1994, Auerbach has accompanied me to Kosrae in part to renew his former ties, in part to make sure that I get the story just right. He has a lot vested in these islands, and cares deeply about the people. His health screening alerted locals to the perils of noncommunicable illnesses and to some possible approaches to preventing them. The screening also confirmed what Ken Miklos has since observed in his dental practice—that the children of Kosrae suffer mightily from vitamin A deficiency, which weakens their immunity and makes them more susceptible to upper respiratory illnesses. The problem would be solved were children to eat the abundant papaya growing everywhere on the island, or the mango. But despite the best efforts of health care workers, they often don't. In Kosrae, papaya and breadfruit are mostly food for pigs.

I am told that among the small victories growing out of the CDC's public health efforts in Kosrae are an early-morning walking program and the Micronesian One Diet Fits All Today (MODFAT) campaign. Through MODFAT, Kosraeans are encouraged to eschew imported food in favor of a diet based on locally grown produce and fish. Auerbach, for one, is wildly enthusiastic about MODFAT. But Dr. Skilling cautions that while it has helped some patients to reduce blood pressure and dependence on diabetes medication, it is not at all certain that MODFAT has had a widespread impact. She suggests that we see for ourselves by attending the MODFAT graduation ceremony, which by grand coincidence is happening just that afternoon.

No ceremony on this island is complete without a feast, and the MODFAT graduates are busy preparing one when we arrive. There are about two dozen women here, flamboyant in billowing muumuus, scooping rice from cauldrons the size of cannons. We cannot help but notice that these MODFAT graduates, so happy and proud, are also obese, some hugely so. We are already late for a flight and beg off lunch, only to find that a delegation of young girls has been sent to our waiting car. Each girl carries a banana leaf heaped high with fresh fish, breadfruit, chicken, chunks of fatty beef in what looks like teriyaki sauce, rice, a can of orange pop, candy, and a slice of thick-crusted apple pie. Apparently the "healthy diet" message has gotten muddled. Later I learn that the walking program is still suspended "for the Christmas holiday"—in May.

Paul Zimmet, an Australian physician and researcher who specializes in the study of noncommunicable disease, wrote in 1996 that "the NIDDM (non-insulin dependent diabetes mellitus) global epidemic is just the tip of a massive social problem now facing developing countries." Zimmet bemoans the "coca-colonization" that has devastated local customs and economies and led to a spread of ill health around the globe. Obesity and diabetes rates have skyrocketed worldwide, but particularly affect traditional cultures in transition—Polynesians, Native Americans, and Aboriginal Australians; Asian Indian emigrants to Fiji, South Africa, and Britain; and Chinese emigrants in Singapore, Taiwan, and Hong Kong. Although the rapid introduction of processed foods and other conveniences is certainly the proximate force behind this trend, scientists are also looking at underlying genetic drivers.

Jeff Friedman and members of his group at Rockefeller University have paid several visits to Kosrae. What they particularly want to learn is why some Kosraeans manage to escape the hazards of "coca-colonization," while others succumb. To Friedman and his team, the interesting question is not why so many sedentary, office-bound, Spam-loving Kosraeans are obese, but why some are not. Friedman's group wants to know what many of us want to know—which genetically linked biological features protect against obesity, and which encourage it. Kosraeans possess a heritage particularly well directed toward answering that question.

Like all natives of Micronesia, Kosraeans trace their ancestry back three thousand years to a handful of Indo-Malayan mariners. Driven by fear, religious persecution, ambition, or ego, this small band of adventurers settled the Pacific. There is of course no written record of the period, but archeological evidence suggests that Marshall Islanders or Melanesians first landed their dugout canoes in Kosrae at the start of the first millennium A.D. Those few who survived this harrowing journey grew into a feudal society composed of three main tribes ruled over by a king and a spiritual leader so powerful it is said he was revered on islands thousands of miles away. Kosraeans committed their share of wife stealing and warmongering, but eventually matured into a rich and complex society of fishermen, priests, and craftsmen.

Ferdinand Magellan, the first European to hazard the Pacific, sailed blithely through Micronesian waters with no clue that there was any "there" there. Two hundred nautical miles from its nearest neighbor, the island went largely unnoticed until 1824, when a French research vessel, the *Coquille,* dropped anchor in Okat harbor. The ship's captain, Louis Isodore Duperrey, and his mate Edmond d'Urville went ashore, apparently surprising a hundred or so feasting natives. Given that until that moment Kosraeans believed themselves to be the only humans on earth, one might have expected hostility. But rather than attacking the white-skinned interlopers, they responded with what d'Urville wrote was "intense joy," offering their visitors platters of breadfruit and small gifts and, in some cases, their women. Rene Primevere Lesson, the ship's "doctor" (he was a pharmacist and botanist by training), described Kosraeans as "advanced people of a high civilization, to judge from the vestiges of customs, tradition such as the authority of the chiefs, classes of society, and the remnants of the arts which they still practice." The women, he wrote, had "black eyes full of fire and a mouth full of superb teeth . . . but a tendency to become fat." The Frenchman also observed that considering the island's bounty, its population of about three thousand was surprisingly small. (At the same time, Easter Island was supporting a population of at least seven thousand with roughly the same land mass and a less hospitable climate.) Studies later substantiated local lore that a much larger population had been whittled down by starvation. Typhoons had devastated the island's food supply.

Once discovered by the French, Kosrae became an increasingly popular base for pirates and for whalers from New England. Both groups brought with them the lethal trinity of tobacco, whisky, and infectious disease. In 1848 and 1849, there were at least sixteen whaling ships in Kosrae harbors, and the following year there were at least twenty-three, some visiting for the second time. Among these was the *Cavalier,* an American ship with one crew member who kept a diary. The sailor wrote shortly after landing in Kosrae: "The females are somewhat modest and do not bestow their favors in the daytime. The starboard watch went ashore at night." A few days later he added this entry: "The islanders are decreasing fast because they smoke too much, drink too much kava, and are too indolent. Sydney vessels brought the pox to the island." The die-off accelerated throughout the nineteenth century. By 1910 fewer than three hundred Kosraeans had survived the Western imports of smallpox, measles, influenza, and sexually transmitted diseases. These three hundred or so survivors comprise the bulk of the ancestry of the island's current population.

James Neel, a geneticist at the University of Michigan Medical School, hypothesized in 1966 that under conditions of feast or famine, natural selection would weed out those unfortunates who could not take in and store food efficiently in their bodies. Neel famously postulated the "thrifty genotype," a constellation of genetic factors that encourages the conversion of calories into body fat and decreases the body's sensitivity to the hormone insulin in order to insure adequate blood glucose levels in the brain during periods of famine. Neel went on to suggest that this mechanism was necessary for survival during periods of extreme stress and scarcity, the so-called bottleneck periods that would otherwise ravage a population. Only those who managed to sufficiently fatten themselves up in advance of the lean times and who could make the best of what was available throughout the period of scarcity could survive, squeeze through the bottleneck, and pass their genes on to the next generation. Hence, in hostile environments like Kosrae, Neel reasoned, the "thrifty gene" was essential to survival.

Societies everywhere have, during some periods in history, lived under conditions of "food stress," and as a result, all of us have evolved some sort of "thrifty gene" mechanism. It is likely, however, that those whose evolution was punctuated by a number of particularly harrowing bottleneck events have developed the most effective versions. In Kosrae, where weather and disease at one time wiped out 90 percent of the population, this effect was likely profound; it is also likely that those who survived these periods had the thriftiest of thrifty genes.

The very genes that once protected these South Sea Islanders and many other indigenous cultures from their punishing prehistory are today predisposing them to early death. Rapid Westernization has rendered their highly efficient genes not protective, but lethal.

In 1994 scientists in Jeff Friedman's group examined blood samples taken in the course of Auerbach's islandwide screening of all 2,286 adult Kosraeans. Preliminary findings suggest that European genes inherited from randy New England whalers and other ethnically European visitors are, in Kosraeans, *protective* against obesity and diabetes. It appears that the more genetically "European" an islander, the less likely he or she is to be obese or diabetic under the current conditions of plenty. Zimmet says this hypothesis is consistent with earlier findings linking Asian genes to those of Native Americans and New Guinea Highlanders. Scientists speculate from these studies that certain aspects of the human genotype thought to have originated in ancient Mongolia have predisposed hundreds of millions—if not billions—of Asians to obesity and diabetes when exposed to the Western diet and way of life.

The stereotypical Asian is thinner, not fatter, than Europeans. But this stereotype is losing power. Asians, too, are getting fat. Worried governments in Malaysia, Indonesia, and Vietnam sponsor anti-obesity campaigns, as does the government of Singapore, where hefty conscripts are required to stay in the military for as many extra months as it takes them to slim down to fighting weight. Obesity rates in urban areas of China have quadrupled in the past decade and nearly one in five Chinese are overweight. Twenty-six million Chinese are diabetic, and 100 million suffer from high blood pressure.

Jared Diamond, a professor of physiology at the University of California Medical School and an expert on ancient cultures, speculated in the journal *Nature* in 1992 that Western industrial nations have already been through a bottleneck of sorts, one that has weeded out some aspects of the thrifty genotype and is keeping diabetes and obesity in the West below the levels now common in Micronesia. "Before modern medicine made diabetes more manageable, " he wrote, "genetically susceptible Europeans would have been gradually eliminated, bringing [diabetes] to its present [relatively] low frequency." Diamond and others have suggested that some human populations, notably those that evolved in Europe, may have developed a relative resistance to noncommunicable disease just as they have to some infectious diseases—through natural selection. Given the burgeoning rates of obesity and diabetes in the United States and other industrialized nations, this seems surprising, until one considers that the rates among the most susceptible—native Hawaiians, Samoans, and Nauruans, some Native American tribes, and Australian and New Zealand and other aboriginal peoples—are higher still. Over half the adults in some of these societies are obese and diabetic, often in response to a level of plenty that is not nearly as high as what is commonly enjoyed in the developed world. Indeed, all measures indicate that the greatest impact of obesity-related disorders will not be in the West, but in newly industrialized and developing nations in Asia, Africa, the Caribbean, Latin America, the Indian and Pacific Oceans, and other regions with historically unstable food supplies.

Genetic susceptibility to obesity once implied a slow metabolism, or as Neel put it, "exceptional efficiency in the utilization of food." But we now know that the overweight generally have a faster than average metabolic rate, and expend not less but more energy than do normal-weight people. In 1998, the year before he died, Neel updated his theory to suggest that the thrifty genotype is better described as a "syndrome of failed genetic homeostasis"—that is, that having thrifty genes has less to do with metabolic rate than with one's inability to self-regulate food intake in the face of plenty.

Andrew Prentice, professor of international nutrition at the London School of Hygiene and Tropical Medicine, agrees that this is probably much closer to the truth. Prentice spent several years working in the Gambia, a swampy, malaria-ridden region where the problem of undernutrition trumps the problem of overnutrition by an impressive margin. In rural regions of the Gambia, farming families drop as much as 50 percent of their body fat during the summer drought. They also lose their fertility. But they survive. For years, Prentice assumed that the Gambians were able to stay alive without food thanks to a slowed metabolic rate. He thought he had verified this hunch by questioning people on their eating habits, and by observing them at mealtimes. From what he could gather, Gambians ate very little or nothing during these periods, and therefore must simply have required less food than did other people, due to a lower metabolic rate. But when Prentice was finally able to get funding and equipment to *measure* food consumption and energy expenditure in the village he studied, he found that the villagers ate lots more—almost twice the calories—than they had reported. As it turns out, Gambians—in particular, Gambian mothers—had the habit of eating on the sly, just as do their Western counterparts.

"You've got millions of people telling you, 'doctor, I just look at a cream bun and I gain a kilogram,' and we were so stupid as to believe them," Prentice says. "The universal impression by doctors and everyone else is that metabolism must underlie weight differences. We've spent hundreds of millions of research dollars looking at various energy-sparing schemes in the body, searching for metabolic effects. And we've roundly failed to find any. To put it bluntly, the thrifty gene might better be called the greedy gene."

Scientists now believe that a genetic inclination toward obesity manifests itself in a constitution more finely tuned to environmental triggers. Primary among these triggers is easy access to high-fat food coupled with a sedentary lifestyle. In Kosrae, as in the rest of the world, genetics was not a problem when the only food available was fish and fruit in the rough—fish that had to be painstakingly pulled from the sea, fruit that had to be harvested. But when polished white rice, cooking oil, fatty meats, and beer became available in grocery stores, most adults on the island grew fat—although not everyone did.

Why some people grow fat and unhealthy in a given environment while others remain slim is the most fundamental question facing obesity scientists. What is clear from studies in Kosrae and elsewhere is that not everyone in a given culture is equally poised to overeat when the opportunity arises. Nor does every culture appear equally susceptible. The French, for example, seem to fiercely resist the pull of adiposity. In 1990 the average French person was just a bit shorter than the average American, but weighed about fifteen pounds less. The French bear a relatively low burden of diabetes and obesity despite enjoying one of the world's finest—and richest—cuisines. This phenomenon is sometimes referred to as the "French paradox," but there is really nothing paradoxical about it. On the whole, the French eat mindfully and in moderation, and at regimented mealtimes. They have a long-standing etiquette and formality built around meals, and do not happily tolerate public gorging. In France an adult loping along with a dripping ice cream cone is considered a figure of fun, and to walk the streets while gobbling a hot dog or sandwich is to risk being tarred as bizarre. "Le snack quick" may be catching on, but it is not a significant part of traditional French culture.

The moralists among us have made a bit too much of French self-control and restraint, as though the French were superior beings from whom the rest of the world could learn a thing or two. Tradition must come from somewhere; the question is, how is it that the French developed a culture of moderation?

There is no certain answer to this, but there are things to consider. France is not a small island. It is a large and fertile country, with a usually reliable and diverse food supply. In 1615, French dramatist and economist Antoine Montchrestien bragged of his country's bounty: "France is alone in being able to do without what she receives from neighboring countries, while her neighbors cannot do without her." France has encountered its share of misfortune: seven nationwide famines in the fifteenth century and thirteen in the sixteenth; the Black Plague in the mid-1300s, wiping out a quarter of the country's population. But as French historian Fernand Braudel has noted, France has more often been "by any standards a privileged country," and no isolated calamity has knocked the country completely off its keel. It is likely that in France there have been fewer decisive "bottleneck" events than in, for example, the tempest-tossed islands of the South Pacific. From this it is fair to speculate that the French of today descend from a people with access to a relatively stable—if often sparse—food supply. In times of plenty, a Frenchman born with a "thrifty genotype" would not have had a true selective advantage, and indeed might have died early of diabetes or heart disease.

This may in part explain why the French, the Swiss, and other western Europeans are less likely than Pacific Islanders to become obese. But again, not too much should be made of this. On a visit to Grenoble a few years ago, I met a chef who complained that the taste buds of his nation's youth were being ruined by an onslaught of salt and sugar processed into an assortment of bland, artless concoctions that dulled their senses and tinkered with their satiety. "When there is no taste, they keep eating," he said. "I worry about my children. Will they, too, become obese?"

The chef is no scientist, nor an historian, but he may have been on to something. Ronald McDonald is no stranger to Parisians. Though still among the slimmest people in the developed world, and slimmer by far than

Americans, nearly one in ten French adults today is overweight. The percentage of overweight French children is higher still.

No culture in the world is immune. The World Health Organization describes overeating as the world's "fastest growing form of malnourishment." While outbreaks of HIV, tuberculosis, and malaria are horrifying, these are not the primary killers of man. That dubious distinction goes to ischemic heart disease (IHD), in which blood flow to the heart is blocked. More than 60 percent of the global burden of IHD is borne by the developing world, and that percentage is expected to grow. The WHO also estimates that the incidence of obesity will double to 300 million cases worldwide by 2005. Most of those affected will live in the developing world, where treatment is hard to come by.

In Kosrae, I meet a twenty-nine-year-old heart attack survivor, a mother of three. Her husband, who is thirty-one, has debilitating diabetes. They are fortunate in that their parents are still alive, to care for their children in case the unthinkable—but far from unlikely—occurs.

In speaking with agricultural and business leaders in Micronesia, it is clear that a concerted government effort to fight noncommunicable disease is not imminent on these islands. I hear repeatedly that health is a matter of willpower and individual effort, and that there is nothing government or business can do to curb the public taste for imports. I hear from their own leaders that the people of Kosrae are "lazy"—too lazy, presumably, to farm or fish for food, too lazy to pound breadfruit rather than open a can.

I've seen breadfruit being peeled with a machete, boiled, and mashed, and I can't imagine why anyone would voluntarily take on this laborious task had they less arduous alternatives. I don't mention this to the business or government people. It doesn't take an expert to know that their charges of "laziness" are a mere distraction. After all, the health administrators and public figures who complain about the lifestyle habits of their countrymen are doing so over a lunch of pork fried rice in a restaurant where breadfruit is not on the menu. The reality is that those in power in Kosrae have a large stake in the commerce that is making their people sick. These power lunchers make light of the fact that many legislators in Micronesia are businessmen, including food wholesalers. They show little interest in discussing the particulars of auto importation in a tiny country already overrun with pickup trucks. Their preference is to focus on the responsibility of individuals to carve out a healthy life in a social and economic climate that makes healthy living all but impossible.

Father Francis Hezel is a Jesuit priest from Buffalo, New York, who has spent more than two decades living, teaching, and writing in Micronesia. White-bearded and springy-limbed, he has the vigor of an evangelical teenager. But he cannot help but be discouraged by what he sees on these islands. "I know that it's a bit odd for a man of the cross to admit, but I don't believe in intentions," he says. "Intentions mean nothing; you have to create structures in which people can perform. The hardest thing to do is to convince people that they can take control of their own lives, and to offer structures to allow them to do this. You can enter any clinic here, and smell the decaying limbs, rotted by diabetes. But even church people here are reluctant to speak out, because they are beholden to the government. They don't want to rock the boat." In Micronesia, as in much of the world, it is more expedient for authorities to promote policies that encourage overconsumption than to discourage it.

Food merchandisers see in the developing world an untapped "land of opportunity." In 1997, according to the World Watch Institute, five new McDonald's restaurants opened every day—four of them outside the United States. "How long can a company of our scope keep doubling its size?" asked Coca-Cola CEO Roberto C. Goizueta. "Where will the next 10 billion unit cases come from? And the 20 billion after that? The fact is that we are just now seriously entering and developing soft drink markets that account for the majority of the world's population, but are also culturally and climatically ripe for significantly increased soft drink consumption." Given this, is it really any wonder why the "climatically ripe" people of tropical Kosrae are no longer content to drink plain water?

Most or perhaps even all of us are susceptible to overeating and to becoming overweight. Our genes tip the balance for each of us at a different point. Some of us succumb easily, others less so. Where we fall in that continuum depends on the match between our genes and the environment they inhabit. Dale Schoeller, a professor of nutritional science at the University of Wisconsin in Madison, is one of many scientists who believe that changes in the environment have pushed a growing number of us to the edge.

Schoeller is something of a legend in scientific circles for creating a method using radioactive "doubly labeled water" to scrupulously calibrate energy expenditure in free-living humans. Before Schoeller, scientists relied almost entirely on questionnaires to measure food intake and exercise—a notoriously inaccurate approach. Schoeller fed volunteers water with radioactive or "labeled" isotopes of hydrogen and oxygen, which allowed him to track the progress of hydrogen and oxygen during metabolism. Some of the labeled oxygen was incorporated into carbon dioxide, a waste product of metabolism, while the rest remained associated with the labeled hydrogen as water. By determining how much oxygen was incorporated into carbon dioxide within a given period of time, Schoeller was able to precisely measure metabolic rate. It was Schoeller who established conclusively that on average, obese people have not a slower but a *faster* metabolic rate than do normal-weight people.

As do many scientists, Schoeller believes that obesity would be a far less prevalent problem were humans to participate in the level of physical activity that nature intended. "Most of the human race regulates dietary intake fairly well when physically active," he says. "But when we stop moving, this regulation starts to fail at the low end. In the last twenty years or so, more and more of us have fallen below a minimum threshold of physical exercise, and we have lost our ability to regulate."

Increasingly, technology conducts the business once performed by our bodies. A parade of labor-saving gizmos—the television remote control, the electric garage door opener, cordless phones, power steering, power windows, food processors—add up to an environment where fewer of us every year are expending the energy necessary to maintain a grip on our appetite. Andrew Prentice says that because it relieves us from finding a stationary phone, using a cordless phone robs us of the equivalent of a ten-kilometer walk each year. Using a television remote rather than getting up to change channels, he reckons, can add up to as much as an extra pound a year.

Appetite control is asymmetrical: our bodies are better designed to respond to hunger than to satiety. When we are inactive, controlling appetite is not instinctive; it is something we have to impose on ourselves. The less we exert ourselves, the more difficult this becomes. "Genetics modulates how quickly each of us falls off the cliff," Prentice says. "But we are all becoming victims of the profound changes in the human environment."

Obesity, diabetes, and other manifestations of the New World Syndrome are not infectious, but like infectious disease, they can be contained with environmental change. In Singapore, the nationwide Trim and Fit Scheme, which began in 1992, has, by regulated school diet and exercise programs, cut childhood obesity by as much as 30 percent, albeit with some measures like a yearly childhood "fatness score" that would be unacceptable in some less authoritarian cultures. In one region of Finland, government-sponsored advertisements, strict food labeling regulations, and school-based fitness programs helped slash rates of coronary heart disease and obesity. And in Hawaii, physician Terry Shintani and colleagues at the Waianae Coast Comprehensive Health Center have shown long-term health benefits from a program emphasizing exercise and a return to traditional local foods.

I spot a glimmer of progress in Micronesia, though not in the organized Western-style wellness programs. One of my hosts, a soft-spoken hospital administrator, confesses that he neither farms nor fishes, but he enjoys playing basketball, and he sometimes jogs rather than drives to the high school gym to play. As a result of this regimen, he has lost thirty pounds, and has managed to avoid some of the health problems suffered by his more sedentary compatriots, including his wife, who is a doctor. Basketball, he says, is catching on in Kosrae, as is baseball. "Imports made us sick," he says. "Now, maybe imports will help us get well."

It is unlikely, though, that other Kosreans will join him on the basketball court soon enough to stem the tide of the epidemic. Most Kosreans would as soon don a pink tutu and tights as jog. When I ask a nurse whether she or her friends or colleagues swim, her lower lip curls into an ironic smirk. Interviews with food importers point to more—not less—processed food being shipped into Kosrae from the West, and the island has just brought in television programming. With television beamed into homes, even making the effort to pick up the night's entertainment at the video shack will no longer be necessary. Kosraeans will be able to return from a day of sitting around a government office, crank open a few cans of Spam, switch on the tube, and kick back for a relaxing evening with the family. It is then that they will truly be able to live—and to die—in the manner of their Western benefactors.

Nine

THE CHILD IS FATHER OF THE MAN

The good is [having] enough to eat and the rest is talk.
—Bertrand Russell

In the waning months of World War II, from September 17, 1944, until liberation on May 5, 1945, famine descended on Holland. The Dutch government-in-exile in London ordered that no trains were to carry supplies to or from Amsterdam, Rotterdam, the Hague, or other cities in the western half of the country, and some 30,000 railroad workers went on strike. The aim was to hinder German troop movements as the Allied Airborne Army landed, an operation that would later be immortalized in the film *A Bridge Too Far.* But this act of pure heroism came at an unspeakable price. The Nazis managed to mitigate the worst effects of the strike by manning the trains with their own personnel. They then retaliated by imposing an embargo on the transport of food to the western cities from the rural north and east. They confiscated bicycles, cars, and the stocks of food wholesalers. They blockaded the roads and the canals. Western Holland was completely cut off and stranded. The hellish months that followed would be known forever as the Dutch Hunger Winter.

November 1944 was miserable, rainy, and cold. There was little fuel for heat, and children were bundled into bed to ward off the bone-deep chill.

City dwellers stripped their homes of burnable wood, then scoured gardens, parks, and the countryside, scrounging frantically for anything flammable. They wrenched out wooden paving blocks from between tram rails, and dredged the canals for rotten timber. In Rotterdam, town squares built on foundations of slag flecked with bits of coal were, as one observer wrote, "turned into mines." But none of this was enough, and hypothermia became the norm. Winter set in and when food supplies dwindled, rations were slashed. By late February, people were issued as few as 500 calories a day, then cut to 400 calories—roughly one-quarter the daily requirement for a small, sedentary woman. A typical day's menu was two small slices of coarse bread, two potatoes, and half a raw sugar beet. Urban folk made weekly treks farther and farther into farm country, swapping family heirlooms for a bottle of milk, a bag of flour, a half-dozen eggs. Some took to eating their beloved tulip bulbs, and thickening watery soup with paper. Wrote one starving man of his fellows, "They only consisted of a stomach and certain instincts."

Generally speaking, starving people cannot produce babies. This is nature's way of being kind, for in hopeless times, the last thing needed is more hungry mouths. Yet, while fertility declined dramatically throughout the Hunger Winter, a surprising number of men managed to maintain their potency, and some women were able to conceive and to carry their pregnancies to term. Pregnant women were theoretically entitled to extra rations, but not all received them, and those who did often shared the few extra morsels with family members. Most lost rather than gained weight throughout the course of their pregnancies, and some died, as did many of their babies. In April 1945, neonatal mortality in western Holland was seven times higher than the wartime norm. Under these conditions, it was something of a miracle that there were any births at all, let alone the thousands of apparently perfectly normal births that occurred. The newborns were, on average, smaller than normal, but those who survived showed no excess of obvious deformity. Learning this, the world breathed a sigh of relief.

Some observers, though, remained leery. The importance of prenatal nutrition had been understood since Napoleon's time, when Pierre Bodin, one of the fathers of perinatology, warned that an army was only as strong as the mothers who bore it. That the Hunger Winter babies could escape unscathed seemed too good to be true. And as time would tell, it was. Hunger would leave its legacy.

It took a generation for the full impact of the Dutch Hunger Winter to begin to be understood. In the 1970s, the husband-and-wife Columbia University research team of Zena Stein and Mervyn Susser, along with epidemiologist Gian-Paolo Ravelli, decided to take another look at the Dutch Hunger newborns, who were by then approaching middle age. They made the rather startling discovery that adult children of mothers exposed to famine during the first two trimesters of their pregnancy were 80 percent more likely to be obese as adults. Equally surprising was that people who had been exposed to famine during their final trimester in the womb or during the first five months of life were 40 percent *less* likely to be obese.

From this, the researchers theorized that deprivation in the first two trimesters primed these famine victims for a life of scarcity. When food became plentiful after the war, this "thrifty fetus" effect backfired, with obesity as the consequence. Later, scientists would also note higher rates of heart disease, diabetes, and other chronic disease and even mental illness in Dutch Hunger Winter babies. These findings have been controversial, and hotly disputed. It is difficult to trace illness back decades to life in the womb, and at first most scientists were staunchly skeptical of these so-called womb effects. But evidence slowly emerged that makes it all but irrefutable that nutrition in the womb and in early life has a very profound and lasting impact on adult health.

David Barker, director of the Medical Resource Council Environmental Epidemiology Unit at the University of Southampton, is almost certainly the world's fiercest champion of the power of life in the womb to shape human destiny. Barker has a reputation as a firebrand, but in look and demeanor he more closely resembles a courtly country physician—the kindly, gracious sort who remembers the first names of his patients' children and grandchildren. We meet in his offices, a two-hour train journey from London in Southampton, a port town perhaps best known for having launched the *Titantic* on its first and final voyage. It is well past sunset by the time Barker polishes off the mess he calls his schedule. He leads us outside through the slop of winter drizzle to his car and home, along the way igniting my flagging attention with insights on the role of pubs in British social life, the Norman conquest of England, and the then-recent marriage of Madonna and British film director Guy Ritchie. "I believe the couple spent part of their honeymoon with one of our neighbors," he says. "Have you heard of him? Sting?"

If David Barker has anything in common with his neighbors, it is not celebrity. By the early 1990s, the Dutch Hunger studies and the theory they spawned were almost forgotten, drowned out by the thunder of excitement over the genetics revolution. Diseases of affluence—obesity, diabetes, heart disease—were considered mostly a matter of DNA and lifestyle: unlucky genes plus careless habits equaled bad health. Barker broke rank by rekindling our collective memory of that cold, desperate winter, arguing that adult health is also determined in the thick amniotic soup and in the first mewling months after birth. Conditions in the womb and in early infancy, he says, "program" the way one's kidneys, liver, pancreas, heart, and brain develop, and how they function later in life. When the fetus strains to accommodate an incommodious maternal environment, or when infants are exposed to malnutrition or infection shortly after birth, permanent changes are hardwired or "programmed" into their brains and other organs, changes that are imperceptible in childhood, but can become dangerous in later life.

We've reached Barker's home, a rambling two-hundred-year-old country manor cradled in acres of sheep pasture. "The connection between lifestyle and disease has been very elusive," Barker begins, uncorking a bottle of white burgundy. "India, for example, has a huge epidemic of coronary heart disease, where it's the commonest cause of death for adults under the age of seventy. Women in southern India don't smoke. They don't drink. They are vegetarians. Yet many are obese and they die of coronary heart disease. How to explain that? To be truthful, there's been a relative bankruptcy of ideas around this. With fetal programming we have something that seems to explain it very clearly."

Fetal programming was for some time derided as fantastical, like something out of Aldous Huxley's *Brave New World,* a mythical land where individual destiny was determined by the broth in which humans were incubated. As it turns out, Huxley was onto something, for womb effects are now widely regarded as critically important. Claude Lenfant, director of the NIH National Heart, Lung, and Blood Institute, says that fetal programming goes far to explain the explosion around the world of chronic disease, particularly heart disease and diabetes. "We know that most diseases result from a mismatch between genes and environment," he says. "The question is, when does the 'environment' piece begin: when you take your first breath of air, or earlier? I say it's earlier, in the womb."

Prenatal programming threatens conventional wisdom by challenging public health doctrine that blames so many bad outcomes on bad behavior. In the late twentieth century, the idea that individual lifestyle choices—in particular, diet—determine health was axiomatic, the foundation for health care policy. We are admonished to cut out one thing after another—sugar, saturated fat, hydrogenated oils, red meat, dairy products, eggs—and told that by taking control over our diet, we can take control of our health. Barker takes issue with this orthodoxy.

"One of the most threatening things about fetal programming," he says, "is that it means that God may not reward you for changing your lifestyle."

To fully understand Barker's thinking, it is first necessary to let go of the cherished myth that mothers-to-be sacrifice all for the developing fetus. The truth is that a pregnant woman's body competes with her fetus for food, and when pickings are slim the fetus sometimes loses out. A malnourished fetus compensates by diverting nutrient-rich blood to its brain and then to its heart, shortchanging its other vital organs. Fetal adaptations are thought to be permanent: The baby is born looking and acting healthy, but its liver, kidney, pancreas, or other organs are subtly compromised in a way that surfaces disastrously later on.

Farmers have long known that restricting the diet of the mother has consequences for her offspring. In adult sheep, nutrient deprivation stunts the fetal liver, making it more difficult for the organ to adequately clear cholesterol from the blood. In adult humans, this translates into heart disease. Likewise, a fetus swimming in glucose inside the womb of an obese, diabetic mother is itself marked for diabetes. Very low-weight newborns are also more likely than normal-weight neonates to become diabetic. These effects help explain why, in just the past decade, rates of adult-onset diabetes—the sort linked to obesity—have skyrocketed around the world, both in the developing world, where pregnant women are so often malnourished, and in the developed world, where pregnant women are increasingly obese. In the United States in 1998, the Centers for Disease Control and Prevention reported a 33 percent increase in the incidence of diabetes between 1990 and 1998. There was a frightening 76 percent rise among people in their thirties. Sources inside the agency say that more recent figures are, as one scientist put it, "even more staggering."

Fetal programming predicts that the risk of heart disease, type 2 diabetes, and other degenerative diseases is acute in hefty adults who were thinner at birth than their genes would predict—the five-pound newborn who should have been an eight-pound newborn and grows into a heavier-than-average adult. Underweight newborns are programmed in the womb for a life of scarcity, and when confronted with a luxurious Western diet, suffer higher-than-average levels of ill health. The theory is bolstered by the devastatingly high rates of chronic disease and obesity in Kosrae and in other Pacific islands, as well as in parts of India, Mexico, and other regions of the world where a sudden uptick in lifestyle collided with a society only recently weaned from poverty.

Studying populations in China, Finland, and India, Barker and others have found evidence that chronic disease spawned in the womb can be passed down through generations, almost like a genetic trait. When an overweight mother overproduces insulin, for example, the hormone crosses through the placenta into the fetus. Too much insulin will tax the fetal pancreas, predisposing the child to diabetes, which she may then pass down to her children. Unless something is done to curtail it, this vicious cycle of ill health can continue indefinitely.

Tufts University Professor Susan Roberts and others have studied children in the shantytowns of São Paulo, Brazil, and concluded that stunting by malnutrition early in life predicts obesity in adults. Deborah Crooks, an anthropologist at the University of Kentucky, found similar effects in poor children living in a rural eastern region of her state. Children stunted from birth by poor maternal nutrition were more likely to be obese, and, Crooks writes, face the same evolutionary nutritional pressures as do children in newly developing countries, like Kosrae. This may help explain why Hispanic and Asian American teenagers born to poor families but brought up in the food-rich environment of the United States are more than twice as likely as their parents to be dangerously overweight.

Barry Levin, a neuroscientist and obesity expert at New Jersey Medical School, writes that "maternal diabetes, obesity and under-nutrition have all been associated with obesity in the offspring of such mothers, especially in genetically predisposed individuals. Altered brain neural circuitry and function often accompanies such obesity. This enhanced obesity may then be passed on to subsequent generations in a feed-forward, upward spiral of increasing body weight across generations." This is merely a hypothesis, but given the evidence so far, it is a highly credible one. Improving infant and maternal nutrition would lower the risk of fetal programming of obesity, and the passing down of "womb effects" to future generations.

"But you can't just hurl food at women in their third trimester of pregnancy and get the desired result," Barker cautions.

Barker's passion comes tinged with irony—the consequence, perhaps, of a professional life eked out on the edge. Educated at Oundle, one of England's most exclusive private schools, he acquired from an early age both a love of natural history and a restless streak. After earning a medical degree at Guy's Hospital in London and a Ph.D. in epidemiology at the University of Birmingham, he and his first wife, Mary, packed up their four young children and moved to Uganda, where he investigated the cause of buruli ulcer, the ghastly condition brought on by *Mycobacterium ulcerans,* a bacterium that spits toxin into tissues, causing swelling, baseball-size ulcers, and, if left to do its work, loss of limbs, eyes, and vital organs.

"It was common knowledge that the bacteria was carried by mosquitoes," Barker says, but he had his doubts. For one thing, there was the niggling fact that the bacteria had not actually been found inside a mosquito. But where else was there to look? The locals told Barker that the sickness came from swamp water. Barker thought that sounded about right. He had mapped the course of buruli in Uganda, and found the disease correlated with proximity to swamplands created by newly flooded regions of the Nile and Lake Victoria. Barker wandered down to the local bog to have a look. A devoted amateur botanist, he recognized patches of *Echinocloa pyranidalis,* a grassy weed notable for harboring all sorts of aquatic bugs, one of which was almost certain to carry the culprit microbe. Unfortunately, Barker had no time to narrow down the suspects. It was 1972, and soon to be "President for Life" Idi Amin Dada was turning Uganda into his own private killing grounds. "We put the dogs down with lethal injections," Barker says. "And we fled."

Barker left Africa convinced that accepting conventional scientific wisdom was not always the shortest path to understanding, a conviction that, a dozen years later, led to the theory that would make his career. He returned to

England and a new job as professor and director of the Medical Research Council (MRC) Epidemiology Unit at the University of Southampton, a rather obscure post where he was left to take on pretty much any project he chose. Mucking about in the bogs of Uganda had alerted him to the relevance of geography to infectious disease, and he was curious to learn whether this might also apply to chronic disease. While looking over a newly edited disease map of Great Britain with his colleague, statistician Clive Osbond, he noticed a striking variation in heart disease rates. With the notable exception of London, men age thirty-five to seventy-four in poor industrial regions of Wales and northern England had substantially higher rates of heart disease than did men in wealthier southern regions. Barker thought this odd, because men in the high-disease regions ate no more fat, smoked no more tobacco, and got at least as much, if not more, exercise than did men elsewhere in Great Britain. Intrigued, he decided to comb through medical records to see what, if anything, in the early life of these men could predict their health status as adults.

"The thing about chronic disease is that it's thirty to fifty years in the making, so to get a clear picture of what was going on in these men, we had to take a look at them as babies," he says.

Barker and his staff scoured archives and hospitals throughout Britain, looking for old maternity and infant welfare records. They found plenty in lofts, boiler rooms, and flooded basements. The records stretched back to the early years of the twentieth century, when the birth rate in Britain was declining and one in ten babies died in their first year of life. Poor health didn't end in childhood: At the start of the Boer War in 1902, the British public read in their daily papers that only 40 percent of British young men were fit to fight—that is, able to run one hundred yards while carrying a rifle. One medical officer wrote that the declining birth rate "must betoken the doom of modern civilization as it did that of Rome and Greece, unless some new moral or physical factor arises to defeat it. It is of national importance that the life of every infant be vigorously conserved." A government committee was organized to investigate and emerged with a Dickensian portrait of the nation's children and mothers: malnourished, ill-clothed, and poorly housed. To improve this dreary state, a public health campaign was launched, and hospitals were asked to keep detailed infant health records. Most of these records were erratic and incomplete, but in Herefordshire, a lush farming county just north of London, the records were immaculate, thanks largely to the sternly efficient Ethel Margaret Burnside, chief health visitor and lady inspector of midwives. Burnside supervised a small army of trained nurses who kept detailed records of every newborn in the county. Barker's team discovered their ledgers in 1986.

With the help of the National Health Service Registry, the researchers located over a thousand Herefordshire babies born between 1920 and 1930, who were by then between sixty and seventy years old. Comparing the infant and adult records, Barker noticed that adults born underweight for their length, or who weighed eighteen pounds or less at age one, were more likely to develop coronary heart disease and stroke than were other infants. It seemed that these people had been marked from birth.

"Interest in developmental plasticity was laid to rest with the genetics revolution," Barker says. "Our showing that it can be linked to disease revitalized the field."

The old model of adult degenerative disease was "based on an interaction between genes and environment," Barker wrote in the *British Medical Journal.* "The new model that is developing will include programming by the environment in fetal and infant life." As it turns out, it is not all small newborns who carry the highest risk of adult disorders, but infants who are smaller than they should be. David Phillips, a member of Barker's group at the MRC, explains that fetal stress can permanently alter hormonal profiles—for example, by throttling up stress hormones to a level that raises blood pressure, heartbeat, and blood sugar. "Nature is about competition," he says. "When the maternal agenda competes with the fetal agenda, there can be some fairly grim outcomes."

Thirty years of animal work documenting the adverse effects of various exposures in pregnancy have made Barker's epidemiological findings all but irrefutable. In 1999 a member of the Southampton group showed that rats born to mothers fed a low-protein diet have high blood pressure. Other scientists have shown that animals deprived of nutrients in the womb grow smaller livers, kidneys, and less flexible blood vessels. "We've found evidence in both sheep and rats that if you produce very mild changes in maternal diet, you can mess up the vascular and endothelial cells, which dictate constriction of the blood vessels and mess up the hypothalamic pituitary axis, which is involved in almost everything," says Mark Hanson, a physiologist and director of the Southampton Centre for the Fetal Origins of Adult Disease. Whether the animal data can be extrapolated to humans is unclear, but Barker is not relying on animal studies to vindicate his ideas. Since 1998 Barker's team has taken the extraordinary step of monitoring the diet, lifestyle, and health of 8,000 Southampton women aged twenty to thirty-four under the not-unreasonable assumption that a goodly percentage will become pregnant before their research funding runs out. At the time of my visit, 730 women had already obliged. Among these is Lynne Allan, a twenty-nine-year-old cartographer in for a checkup. Eight months pregnant, Lynne is remarkably good-humored throughout the nearly two hours she is monitored, measured, and interrogated down to the brand of butter she spreads on her morning bun. While not a few Southampton women seem to subsist on tea, fried fish fingers, cigarettes, and the not-so-occasional gin, Lynne is a vegetarian who admits sheepishly to a taste for chocolate biscuits.

"Since you can't experiment on people, you have to observe them," says study coordinator Hazel Inskip. "But epidemiology is such a weak science, we're forced to follow this large group and ask so many questions . . . it's using a sledgehammer to crack a nut."

By monitoring women's health before, during, and after pregnancy, and following their children, the Southampton group hopes to tease out precisely what factors in the mother's diet affect fetal development and child health. "What we really want to find is a way to effect change," says Nancy Law, a physician in Barker's group. The changes the team has in mind focus on promotion of health rather than merely treatment of disease. Barker has little patience for those who refuse to accept what he considers to be the patently obvious fact that health has its genesis in the womb. American epidemiology, he says "has gone off on a boil, to rooms filled with paper and no patients." He finds particularly vexing pronouncements emanating from the famous Nurses' Health Study, conducted at the Harvard School of Public Health, almost certainly the most influential long-term epidemiological study in history. The Harvard effort is known for its promotion of the the so-called Mediterranean diet, a meal plan built around olive oil, whole grains, and vegetables that is said to reduce obesity and chronic illness. Barker considers this bunk.

"The Greeks eat a Mediterranean diet and they have one of the highest rates of diabetes (and obesity) in the world," he says. "You can't consider nutrition in isolation, can't say, for example, that the Mediterranean diet is good for everyone. Each of us has an Uncle Charlie who lived a grand life, drank and ate up a storm, and died at age 100, while his brother Uncle Frank lived a blameless life and died of coronary heart disease at age forty-five. But we can't blame everything on genes and go to sleep, because genes don't quite explain it either. People have to open their mind to the terrible possibility that this theory might be right—that the important events in the development of a child begin not at conception, but with the health of the mother."

In 1999 Anita Ravelli (no relation to Gian-Paolo) published a study she authored with Barker and others that compared the by-then middle-aged survivors of the Dutch Hunger Winter with another group. She found that those who were exposed to famine early in their gestation were more likely than others to be fat, while those exposed to famine later in gestation were not. The authors conclude that "fetal life seems to be a critical period for the development of obesity," which they ascribed to "permanent adaptation of central regulatory mechanisms of energy intake and expenditure." Matthew Gillman, associate professor in the department of ambulatory care and prevention at Harvard Medical School, says the relationship between fetal life and obesity is complicated. People born at a very high weight, he explains, are generally fatter than average at middle age. But very low birth weight infants who later get fat are in even greater danger, due to their tendency toward central obesity, the thickening around the middle linked to heart disease, diabetes, and high blood pressure.

A predisposition toward obesity may be "written in the genes" but it is not indelible. The inclination is to some degree alterable during various periods of life: prenatally, as Barker has shown, but also in early childhood and into adolescence. As children age, this window of plasticity starts to close. A fat but otherwise healthy two-year-old is no more likely than other babies to grow up fat, but an obese fifteen-year-old has about an 80 percent chance of being obese as an adult.

A generation ago, pediatric and adolescent obesity were rare enough that they were not surveyed in the population. Adult standards—set by the body mass index—didn't seem to apply to young and growing bodies, and it was assumed that chubby children and teens would outgrow their baby fat. But as it became increasingly clear that obesity posed dangers for children, the federal government set body weight standards adjusted for them. These standards, though far from perfect, offer evidence of the extent of the problem.

Childhood obesity in the United States jumped from 5 percent in 1964 to 14 percent in 1999, the last year for which survey information is available. In Australia, childhood obesity rates tripled between 1985 and 1995, and today one out of every five children there is overweight. Obesity-linked "adult onset" diabetes mellitus is for the first time being reported in children and adolescents in Canada, Japan, India, Hong Kong, Singapore,

Bangladesh, Libya, the United Kingdom, and New Zealand, among other countries. Equally alarming are growing numbers of reports of "super-obese" children, of 120-pound three-year-olds, and of 400-, 500-, and, even 600-pound teenagers.

Limited but persuasive evidence points to morbid obesity as a factor in early death in children. A visit to the endocrinology unit of any major children's hospital offers a painful glimpse at the sort of life these youngsters are leading, of the anguished indignity suffered by a ten-year-old confined by his bulk to a wheelchair. And it is certain that even less severely overweight children carry risks that can haunt them into adulthood. Biological indicators are clear—high blood glucose and lipid levels and elevated blood pressure among them. These are all precursors to adult heart disease.

In 1985 Steven Gortmaker, a psychologist and an associate professor at Harvard Medical School, and William Dietz, a pediatrician who now heads the nutrition and physical activity division of the Centers for Disease Control and Prevention, collaborated on a landmark study of obesity and television viewing. They found a clear association between the number of hours of television a child watched and the risk of that child becoming obese or overweight. In twelve- to seventeen-year-olds, the prevalence of obesity increased by 2 percent for every hour of weekly tube time. A more recent study found that while 8 percent of children watching one hour or less of television a day were obese, 18 percent of children watching four or more hours were obese. Another study concluded that children who watched five or more hours a day were more than eight times as likely to be overweight as those watching two hours or less. The problem is not, of course, confined to the United States; in Australia, Japan, and much of the industrialized world, television viewing has been implicated as a serious contributor to children becoming overweight.

The more television children watch, the more they eat, which seems odd when one considers that watching television does nothing to rev the metabolism. (By comparison, even reading is a workout, at least in studies that have been done with obese children, perhaps because it engages their minds a bit more emphatically.) Television viewing prompts children to consume more food while they consume less energy, an ideal recipe for adiposity. And by far the foods most commonly peddled to kids through television advertising are soft drinks, candy bars, and sugary breakfast cereals.

It is no secret that most high schools and many middle and even elementary schools stock vending machines with soda and other treats. In many cases, these are available to students throughout the school day. School cafeterias have also given themselves over to fast food outlets. Perhaps like most parents, I hadn't given any of this much thought until my younger daughter entered middle school in the fall of 2001. She received a lunch menu in the mail—and couldn't believe her good fortune. The options included chicken nuggets, fried chicken patty with French fries, Domino's pizza, cheeseburgers, Pop-Tarts, and Rice Krispie Treats. To drink there was soda, fruit punch, and chocolate- or strawberry-flavored milk. The cookies and ice cream bars were particularly recommended. So much to choose from! And the kicker was this: now that she was a sixth grader, and very busy, there would be no time for recess. And due to budget constraints and a demanding academic curriculum, physical education class would consume no more than an hour a week of her valuable time.

For decades, Americans have waxed hot and cold in our support of physical education in the public schools. In the years following the Great Depression, school fitness programs were slashed along with art, music, and other "nonessentials," victims of austerity education budgets. But when a 1944 survey by the Office of Education revealed that only half of all high school juniors and seniors were enrolled in a physical education course, concern rose. This was wartime, and the possibility of recruiting high school age students into the military was very real. Fit adolescents, it was reasoned, would make better soldiers. After all, the Soviet Union had thrown itself full force behind a youth "physical culture" movement, at least rhetorically.

In *I Want to Be Like Stalin,* their 1947 text outlining principles and goals of the Soviet educational system, propagandists B. P. Ypsipon and N. K. Goncharov wrote that "physical education as a whole promotes the development of those qualities which are essential to future warriors of the Red Army."

Americans worried they were falling behind. Cold War paranoia equated physical softness with strategic vulnerability, and fitness gradually became a national security issue. In the 1950s, the Korean War uncovered a weakness in the American military—a surprising number of soldiers suffered from debilitating back pain. Hans Kraus and Sonya Weber, physicians in the Posture and Therapeutic Exercise Clinic of Columbia University Vanderbilt Clinics, spent years studying the problem, and after examining thousands of subjects concluded that 80 percent were what they called "weak-muscled." Wondering at what stage in development this weakness surfaced, they tested four thousand schoolchildren ages six through sixteen in cities and towns along the Atlantic coast. Nearly 60 percent of the children failed the test. This finding was so dramatic that Weber and Kraus decided to see whether the problem was an artifact of youth—that is, whether most children were simply, by definition, weak. They turned to Europe for a comparison group, and tested two thousand children in Austria, Italy, and Switzerland. Only 8.2 percent of European youngsters showed weakness. The scientists concluded that American children, though among the best fed and healthiest in the world, were woefully underexercised.

This rather shocking revelation was widely publicized. Hans Kraus himself went on the lecture circuit, and eventually ended up at lunch with President Dwight Eisenhower. Eisenhower played golf, watched his weight, and considered himself a pretty fit guy. He took Kraus's comments seriously, and showed his concern by appointing Vice President Richard Nixon as head of a committee to design a national fitness program. A year later, in 1956, the President's Council on Physical Fitness was organized to rally a national fitness campaign. The council urged that legislation be enacted at the federal, state, and local levels to tackle the growing problem of childhood sloth. This did not happen. *Sports Illustrated* described the effort as "primarily one of vaguely worded publicity and promotion releases for the fitness cause." In 1958 Vice President Nixon relinquished his chairmanship of the council to Secretary of Interior Fred Seaton. Seaton opined publicly that the best way to become fit was to take up bird watching.

A couple of years later, John F. Kennedy took a somewhat more strenuous position. In "The Soft American," an essay for *Sports Illustrated* published in 1960, he wrote: "The President and all departments of government must make it clearly understood that the promotion of sports participation and physical fitness is a basic and continuing policy of the United States." In this, Kennedy was a man of his word. Among his earliest acts as president was to appoint a committee of high-profile sports figures to determine what should be done about the sagging American physique. The council recommended that school children be required to attend at least a half-hour a day of physical education, and that they be permitted a substantial recess. This advice was eagerly and earnestly received, and largely ignored.

Education reforms in the 1960s and 1970s cast doubt on whether a prescribed course of physical education was necessary, or even equitable. Support for physical education waffled, and then, gradually, phys ed regained its old, Depression-era stigma, increasingly regarded as a luxury that schools could no longer afford, in particular given the urgent push to raise academic test scores. By the late 1990s, physical education had become a dangerously low priority, with many school systems—and most states—abandoning the requirement altogether. The Centers for Disease Control and Prevention warned in 1999 that fewer than half of high school students were enrolled in physical education classes for even one hour a week, and an article the following year in the journal *Pediatrics* was even bleaker. It concluded: "Despite the marked and significant impact of participation in school PE programs on physical activity patterns of U.S. adolescents, few adolescents participated in such school programs." The authors went on to make the rather extraordinary observation: "In addition to the more readily modifiable factors, high crime level was significantly associated with a decrease in weekly moderate to vigorous physical activity."

Many American children do participate in sports: there is a growing proliferation of soccer, basketball, and baseball leagues, gymnastic clubs, tennis teams, karate, judo, and dance classes. But increasingly these are elite, exclusionary opportunities requiring parental involvement and funds. Gifted athletes are always welcome, but the less than gifted, particularly those of lower incomes—are likely to drop out of organized sports and sports clubs before entering their teens.

Steven Gortmaker has studied childhood obesity for nearly two decades, and over that time, he has developed some rather strong views. When we talk he fixates on the ceiling from time to time, as though gathering patience to address an issue he has addressed so many times, and to so many deaf ears. "We have a multibillion-dollar food industry whose goal is to get people to increase their food consumption as much as possible," he says. "They've been very successful. Every public building has a soda machine, every community, no matter how small, has drive-through restaurants serving ever larger portions of food at lower and lower prices. We have a multibillion-dollar video game, computer, and television industry whose goal is to get more and more people viewing screens. So what we have now is an imbalance of caloric intake and energy expenditure, driven by these huge industries that reinforce each other through advertising. They've put out the idea that food is fun, that food is exciting, and they've attached to the food other exciting things like action figures and cartoon heroes. Just think about it—by increasing the body weight of the American population, these industries are quite literally expanding their markets. It's a gold mine."

TEN

AN ARM'S REACH FROM DESIRE

> We may not pay Satan reverence, for that would be indiscreet, but we can at least respect his talents.
>
> —Mark Twain

The Food and Beverage Marketing Conference is barely under way when speaker Amanda Smith, senior consumer research specialist at International Flavors and Fragrances, cuts to the bloody quick. What, she asks, is kids' number one flavor? The hundred-odd marketing and advertising specialists leap to the challenge.

"Chocolate chip mint!" No.

"Cookie-dough!" No.

"It's gotta be creamsicle!" "Cherry!" "Watermelon." ". . . Kiwi?" No, no and no.

Smith smiles a Sphinx smile, sips from her bottle of water.

"The number one flavor," she says, "is blue raspberry."

The audience groans. They should have known!

Raspberries are red, of course, or black. Not blue. But if you're hung up on that you've missed the point. It's not the raspberry that's hot, it's the blue—the *color* blue. Green is hot, too, which explains green ketchup. But at the moment green is not as hot as blue. Blue may be the ketchup color of the future, or purple. But that is for the customers to decide.

Offering "control" over "choices" is the key to food marketing. We get to "choose" the color of our ketchup. We "choose" whether to have chocolate or cinnamon breakfast cereal. We "choose" to dunk our chicken nuggets in honey mustard or barbecue sauce. Choice is what a free society is all about, and these marketers and advertisers offer all the choices we customers can handle.

Or rather, the *perception* of choice.

Kid Power Exchange, described in promotional literature as a "global resource that assists organizations in optimizing their youth marketing goals," runs an annual meeting for professionals who make their living making, marketing, and selling food aimed at the prepubescent set. This year's event takes place in a bland function room in a flavorless hotel in Atlanta. Featured talks include "Making Emotional Connections and Building Relationships with Kids" and "Creating New Food Brands for Today's Kids . . . Truths and Consequences." Food marketers pay $1,699 to attend this two-day meeting because kids have money, sometimes big money, and spend the largest portion of it on candy, soda, and salty snacks. But kids' buying power extends well beyond their personal means. When it comes to food, the youth market is, to large degree, *the* market.

James U. McNeal, professor emeritus of marketing at Texas A&M University, is a world authority on selling to children. He has made a career elaborating on the truism that while parents may have the keys to the family car, children have the key to the family food budget. The trick to food marketing, he contends, is to enlist children as lobbyists. In *Kids as Customers*, McNeal unveils a catalogue of horrors he calls the seven varieties of "pester power," a set of persuasive tricks used by children to get parents to buy what they otherwise would not buy. Among these are the *persistent* nag, who begs, pleads, and grovels; the *pity* nag, who rationalizes that the acquisition of certain objects is prerequisite for peer acceptance; and the *demonstrative* nag, who threatens tantrums or worse if his or her every wish is not granted. According to McNeal, 75 percent of spontaneous food purchases can be traced to a nagging child. And one out of two mothers will buy a food simply because her child requests it. To trigger desire in a child is to trigger desire in the whole family.

"Kids' food and beverage products should not be screamingly 'kid,'" warns one speaker at the Atlanta conference. "If you go too far, Mom will resist. Food products just need to be fun. Roughly speaking, consider your target 80 percent the kid, 20 percent the mom. And when your ad dollars are limited, go to the kid alone. When you are trying to reach a family, kids are the primary target."

In his most recent book, *The Kids Market: Myths and Realities,* McNeal outlines a theoretical mother-and-child supermarket trip in which an eight-year-old who has already mastered every imaginable strategy of persuasion thwarts her mother's good intentions. In a typical exchange, Mom asks, "What kind of lunch meat do you want for your sandwiches next week?" The kid responds, "I don't want lunch meat, I want Lunchables." The mother may have never heard of Lunchables, a tray of bite-sized snacks marketed by Kraft as a complete meal. But in this scenario, the harried woman tosses Lunchables into her shopping cart and hurries on. In total, the tyke "scores" two dozen such requests, which McNeal describes as "a relatively small number," one that would be much larger were two or more siblings in tow. (One might extrapolate from this McNeal's ideal customer: an exhausted working mother making her supermarket rounds at 6:30 P.M. on a weeknight, trailed by three or four frantically hungry tots.) He concludes that "fulfilling the requests of kids in such a way that the kids are happy and healthy brings parents and children closer together in a more loving relationship. All parents want these loving relationships with their kids, and they [the kids] want them, too. If deferring to kids helps produce it, so much the better."

In this world order, the family consists of one or two working parents too busy to cook, too busy to shop, and too frazzled to exert authority over the family's eating decisions. And love is a matter of deferring to the taste preferences of hungry, television-addled five-year-olds. Preparedfoods.com, an on-line newsletter, reports that "kids decide what they eat in half of their meal occasions." Leaving open the question of how meals of chocolate cereal straight from the box qualify as "occasions," the newsletter goes on to note the rising popularity of "parallel pantries," where parents maintain a separate stash of favorite foods for the little ones in order to minimize "dinner-time battles." Preparedfoods.com suggests that creative marketers "capture younger millennials" (a.k.a. children and grandchildren of baby boomers) with "tools for fun. Contrasting and complex yet familiar flavors, wild colors, magic, new forms, shapes, and innovative packaging."

Given the overabundance of food in the industrialized world, simple nourishment is not the focus of food marketing. We have more food than we need—much more—and the food industry's job is to sell sizzle long after we have had our fill of steak. After all, 116,000 new "food products" were introduced to the U.S. market between 1990 and 1998, each one designed for consumer appeal. Like the tobacco industry, food marketers peddle image, and like the tobacco industry, they prefer to reel in customers while they are young. Generally speaking, taste in food—like brand loyalty in cigarettes—is cultivated in childhood.

Karen Picciano, an account director for Alcone Marketing, has come to the Atlanta meeting to speak on behalf of Burger King's Kids Club. The Kids Club target is "kids, and the families that they bring with them for a dining experience." The average check for a family is three times that of a "non-family transaction," so young children with parents in tow are especially courted at fast food chains. Kids Club won't induct kids before age three, but mothers can snare a spot for their newborn on the Kids Club waiting list. "From age zero to three we're talking to the mom," Picciano tells the audience. "This gives us access to an unparalleled knowledge base. We know who these kids are and where they live and it allows us to align this information with other marketers." To retain loyalty, Picciano says that Burger King "completely caters to kids," offering "instant gratification, like new toys that are constantly cool, toys that reinforce the brand and also drive traffic through the door, with kids who want to collect the whole set." At lunch that day I sit down with two executives whose companies make and sell these toys. They tell me that their toys are not cheap, that in fact that the cost of these "premiums" can sometimes blot out profits from the sale of the kid's meal. But the fast food industry's goal in luring kids is to lock in lifelong customers, not necessarily to turn a quick buck.

Worldwide there are five and a half million active Kids Club members, and more than 6 million graduates. The company processes between twenty and thirty thousand new Kids Club applications each week. Picciano outlines the themes of Kids Club as the "themes of childhood": security and belonging (a club is "exclusive"), separation and freedom from adults (no one older than twelve allowed), mastery (Kids Club games, toys, and humor are dumbed down so that every kid can feel smart), and most important, the power to make choices. "For example," she says, "to decide they want no onions or tomatoes on their hamburger."

But Burger King customers cannot choose to have their burger medium rare because the grilling time is regulated. They can't choose to order deli favorites like a turkey sandwich on rye or a bowl of chicken noodle soup or even a sour dill pickle because Burger King doesn't offer those things. To be profitable, convenience foods require economies of scale, and to achieve these, choices are offered only within an extremely narrow realm.

Political scientist Benjamin R. Barber writes in his consideration of globalization, *Jihad vs. McWorld,* "This politics of commodity offers a superficial expansion of options within a determined frame in return for surrendering the right to determine the frame. It offers the feel of freedom while diminishing the range of options and the power to affect the larger world." That is, we can choose to have the pickle left off our burger, but we can't choose to have a glass of tomato juice or a fruit salad.

The nearly $112 billion American fast food industry claims it would be delighted to offer healthy options, but that, sadly, consumers—fools that we are—just don't buy them. Not long ago, Wendy's restaurants stuffed carrots, red cabbage, cucumbers, red onion, broccoli, and romaine into a pita pocket and sold it as a sandwich. Some people liked it—liked it a lot—and Wendy's garnered kudos from consumer interest groups and health advocates. But after a few years, Wendy's unceremoniously dropped Pita Pockets from its menu, claiming lack of customer interest. Taco Bell's Border Lights suffered a similar fate. The low-fat line was lavishly launched in a $20 million dollar marketing campaign in which eight million tacos and burritos were given away free. But the company claims that the very presence of a low-fat option alienated regular patrons who mistakenly assumed that Taco Bell's entire menu had gone healthy. Overall sales slipped, and stockholders pressured the company to drop the low-fat line, which Taco Bell did with little fanfare.

An even greater disappointment was McDonald's McLean Deluxe, a sandwich designed to serve two masters that spectacularly failed both. The Deluxe was the brainchild of food scientists at Auburn University in Alabama, who in the late 1980s were looking to create a lean hamburger that would be as tasty as the fatty fast food standard. Fast food hamburger patties are 20 percent fat, 20 percent protein, and 60 percent water. Remove the fat and the burgers become dry. The problem was to find a way to hold in moisture while losing the fat. The solution was carrageenan, a seaweed derivative that binds water. The Auburn team traded off ten grams of fat in ordinary ground beef for water, flavoring, and a smidgen of carrageenan. The burger fried up nicely. They dubbed it the "AU Lean." In blind taste tests, one hundred families agreed that the AU Lean trumped the fat burger for "flavorfulness," "juiciness," and "likability." In 1990, the Auburn group convinced McDonald's to try AU Lean in its franchises, and in 1991 McDonald's put the McLean Deluxe on the menu in 1,600 markets. It had 340 calories and a scant 12 grams of fat. George Rosenbaum, president of Leo J. Shapiro & Associates, a Chicago market research firm, said at the time that the burger could lead to a whole menu of low-fat, low-calorie foods, just as the Egg McMuffin had led to a whole new breakfast menu.

The McLean did not live up to its promise. Comedian Jay Leno joked that the burger was a huge hit with "all four" McDonald's customers interested in nutrition. Columnist Mike Royko wrote: "It wasn't the worst thing I ever ate. Some years ago, while fishing in the Ozarks, I yawned and a large bug with big wings flew into my mouth. That was really disagreeable. On the other hand, dressed up with onions and ketchup, the bug might have proved a better snack than the McLean thing." Half a decade later, the McLean was history, replaced by the Arch Deluxe, the "adult" hamburger with 560 calories and 32 grams of fat—more with the optional bacon layer.

If those one hundred Alabama families are to be believed, though, the McLean, when properly cooked, apparently tasted just fine, even better than the greasy original. But regulation fast food practice doesn't allow for careful cooking—so the McLean was generally overcooked into a hockey puck. (Some franchises so resented bothering with the things that they fried up normal burgers and pawned them off as McLeans.)

That said, it wasn't just taste that killed the McLean: it was the concept. Conspiracy-minded health advocates charge that McDonald's itself poisoned the well, and there is reason to take this suggestion seriously. For reasons that are unclear, McLean was priced slightly higher than the heftier Big Mac, giving the impression of less value. It was promoted as "lean" and "healthy" rather than "tasty," not a great marketing strategy for a fast food. (Even Weight Watchers trumpets its reduced-calorie food lines as "indulgent" and "decadent.") And McDonald's, like Taco Bell, may well have come to the gradual realization that putting a lean burger on its menu was having an unintended effect on total sales. The lower-fat burger aroused calorie consciousness, inadvertently alerting customers to the downside of extra-large orders of fries and Cokes. Given that French fries and soda cut fast food chains their thickest profits, McDonald's may well have been relieved to retire the McLean, just as Taco Bell might have been glad to say *adios* to Border Lights.

Five years before introducing the McLean, McDonald's launched a major advertising campaign in the United States aimed at neutralizing what it called "the junk food misconceptions about McDonald's good food." An internal public relations memo advised: "McDonald's should attempt to deflect the basic negative thrust of our critics. . . . How do we do this? By talking 'moderation and balance.' We can't really address or defend nutrition. We don't sell nutrition, and people don't come to McDonald's for nutrition."

The *American Heritage Dictionary of the English Language* defines nutrition as: "The process of nourishing or being nourished; especially the interrelated steps by which a living organism assimilates food and uses it for growth." If McDonald's is not about nutrition, it's fair to ask what it *is* about. It is in the business of selling food, which, for most of us, at least implies nourishment. But the fast food industry has managed to disassociate food from its primary purpose. The food it sells is meant to divert us. That many of these foods offer little in the way of actual nutrition seems to matter not a whit.

Raise this reservation with food industry executives, though, and they'll break into a well-rehearsed refrain: "All foods can be part of a healthy diet." Other retorts include, "We advocate that people eat a variety of foods" and "There is no such thing as a bad or a good food." Moderation, they intone, is key. In this they are echoed by a Greek chorus of industry-backed "consumer awareness" groups. Primary among these is the American Dietetic Association (ADA). Frequently quoted as an "impartial" source in media reports, the ADA has "partnerships" with food lobbies such as the Dairy Council and the Sugar Association, as well as individual food companies. Another industry-sponsored group, the International Foods Information Council (IFIC), supports "media awareness" around nutrition issues, in part by paying journalists to serve on advisory committees and to lead "nutrition awareness" workshops. The Institute of Medicine Food and Nutrition Board is sustained in part by M&M Mars, and the American Society for Clinical Nutrition by Bestfoods and Coca-Cola. The message these and other arms of the food industry lobby so hard to get across is that the right to make one's own food choices is sacred, and that to argue otherwise constitutes a Big Brother-style breach of civil liberty. What is overlooked in these arguments is that we no longer have the choice of ordering a McLean.

At the Atlanta conference, I meet for a drink with a consultant, one of the one hundred or so who help craft Burger King's marketing strategy. She looks to be in her early forties, and every inch the "soccer mom with an attitude" type whom food marketers strive to cultivate. She admits that a meal consisting of a "variety" of Burger King offerings is not ideal, and has particular misgivings about the "Big Kids' Meal"—an engorged version of the regular kids' meal containing more than a thousand calories and fifty grams of fat. She says she would never wish such a thing on her own children. But she couldn't do her job if she didn't believe that Burger King legitimately served the needs of some constituency.

"You can't overlook roofers," she says. "Roofers work really hard, and they need a lot of calories."

One can't help but wonder why she didn't add "marathon runners" or "lumberjacks" or "bicycle messengers" to the list. Some people in the industrialized world do require a lot of calories, but not many. In truth, Burger King and other successful fast food chains do not cater to the tastes of roofers, or, for that matter, adults of any profession. Fast food, like most convenience food (with the possible exception of beef jerky), is kids' stuff: soft and sweet, or salty and bland. Fast food calories are built from fat, protein, and simple starch. Fast food chains and quick food outlets, like gas stations and convenience stores, minimize the use of perishable foods, which cuts out most fresh vegetables and pretty much all fresh fruit. This means that quick food usually offers little in the way of micronutrients or fiber. Iceberg lettuce and translucent slices of tomato and onion are the most common and often the only fresh items.

"We have salad," the Burger King marketer says. "We sell about two a week. We offer it to avoid the veto factor. That's when people go out to lunch with a group from the office and one is on a diet. If we didn't have salad we'd lose the whole group. But we don't focus on salad and we don't promote salad because it's not a money-maker. It goes bad too quickly. Take a look at the lettuce at the back of your refrigerator and you'll see what I mean."

The word "restaurant" derives from the French *restaurer,* to restore, but there is little opportunity for that in fast food emporiums, which make a sort of silent deal with consumers: they feed us cheap and fast and we carry our own trays, chow down, clean up, and get out. Teenagers and senior citizens who willingly work for minimum wage and no benefits are desirable fast food employees, but they are not courted as customers because they tend to linger a bit too long over beverages and shared orders of fries. The time spent by an average cusomer in a fast food restaurant is a blistering eleven minutes. Stiff plastic chairs bolted to the floor a tad too far back from the table encourage us to eat and run, but usually, the chairs are beside the point. Most of us don't even bother to sit down.

The introduction thirty years ago of the drive-through window made possible the purchasing of meals from the comfort and privacy of the car. Today 60 percent of fast food sold in the United States passes through the drive-through, and menus and packaging are changing to accommodate the trend. NPD Group, a Chicago-based tracking firm, estimated in the mid-1990s that one out of ten fast food meals purchased was consumed in the car. That figure is almost certainly higher today, and food purveyors seem thrilled. Church's Chicken is experimenting with a flip-top French fry carton that fits snugly into car cup holders, making it easier for drivers to dip fries or "Big Muncher Chicken Strips" in sauce while navigating the turnpike. Kentucky Fried Chicken's "Twister" sandwich—fried chicken wrapped in a tortilla to prevent dripping—was created with the same thing in mind. Breakfast sandwiches—breakfast meat, egg, and cheese on a biscuit, bagel, or croissant—were designed to be eaten on the run or at the desk. Meals are blurred in a streak of what the food industry calls "feedings," six or seven daily pit stops described variously as "deskfast" and "carfast" and "dashboard break." "The big joke is that the next development will be a feed bag, so that people can snack all day rather than sit down to a decent meal," the Burger King marketing consultant says.

Convenience is more than a buzzword in the food industry; it's a matter of life or death. Millions of advertising dollars are spent reminding us that we do not have the time or patience to shop for, cook, or thoughtfully consume our meals. And we are convinced. In the 1960s homemakers spent about two and a half hours making dinner each night. In 1996, the latest year for which figures are available, dinner preparation had shrunk to fifteen minutes. Technology has made possible devices and products, such as frozen food and microwave ovens, designed specifically to save us time and energy. Expectations set by these products have made us even more impatient.

"Americans can't abide slowness," cultural historian David Shi told *The Washington Post.* "Waiting has become an intolerable circumstance." Convenience foods have "evolved" into such space-age concoctions as "tube" foods like Berry Blue Blast Go-Gurt, which is squeezed directly into the mouth. These products seem not to be food for active youths or healthy adults, but pablum for toothless invalids. They are not eaten, really, but sucked, as one would suck a thumb. And that's just the point, for to eat in such a distracted, habitual manner is to eat more than one would mindfully eat. The industry has a name for food eaten this way: "no-think foods," that by definition "don't drip, crumble, require utensils, or demand inordinate attention."

"No-think foods" are the hottest products in new product shows: the appetizers we wolf down while we're waiting for the main course to arrive, the salty snacks we munch while we're watching television, the greasy sack of chicken nuggets we nibble furtively while we're driving home from a hard day at the office. The food marketer's goal is to convince us to eat these things without noticing—so that we will eat again.

The National Restaurant Association ranks in *Fortune* magazine's list of the twenty-five most powerful lobbies in Washington, and the organization lobbies tirelessly against regulation or taxation in the food industry. It also wages a forceful propaganda campaign against consumer interest groups, which warn against the calorie, fat, and sugar content of restaurant offerings. The charge is that these scolding "food nannies" or "fat police" are attempting to limit the public's right to choose, and that the imposition of taxes on soft drinks, restaurant meals, or snack foods will somehow lead to the "elimination" of these foods. (California's 7.25 percent sales tax seems to be a particular annoyance.) The organization is keenly sensitive to criticism over portion sizes or fat content in restaurant food, precisely because serving large portions of high-fat food seems to be the key to success in the restaurant business. In 1999 Americans averaged 139 restaurant meals per person, a 14-percent increase over the previous decade. Eighty percent of that increase was captured by fast food outlets.

Reporting on the success of fast food chains in Western Europe, one industry newsletter reports that "the power of the large chains is evident, with every country recording a decline in the proportion of value sales through independent units. Chained operations expanded strongly in Western Europe over the review period, including multinationals, such as McDonald's and Burger King, and regionally focused chains, such as TelePizza and Autogrill. The economies of scale enjoyed by the chains, in terms of cost control and ability to locate in prime sites, tends to give them higher turnovers per unit than independents, thus boosting share of the market in terms of value sales."

Social critic Benjamin Barber and others have charged that McDonald's is "making war" on the European tradition of mealtime, but that is not entirely fair. With more women entering the workforce and less flexible working hours, many Europeans face time pressures that make a leisurely family lunch impractical, if not impossible. What the French, the Italians, and much of the world truly fear is not the loss of a two-hour lunch, but the elevation of fast food style instant gratification above all other concerns. In an editorial titled "Vive le Roquefort libre!" *Le Monde* warned that "McDonald's red and yellow ensign is the new version of America's star-spangled banner . . . whose cultural hegemony insidiously ruins alimentary behavior." *Le Monde* has reason to worry. In *Fast Food Nation,* journalist Eric Schlosser describes fast food as a revolutionary force that has transformed not only the American diet, but also our landscape, economy, and popular culture. He makes a persuasive case. According to Schlosser, one-quarter of American adults eat fast food on any given day. More than a few eat fast food twenty times or more a week, which is to say, almost three times a day. These "heavy users" are typically traveling salespeople eating in their cars, or young unmarried men who either don't know how or won't bother to cook.

A few years ago, *The Wall Street Journal* ran an eye-opening story detailing the exploits of a group of "heavy users"—friends who worked together in a furniture moving business. All were seriously overweight and one, a twenty-seven-year-old former high school football star named Phil, was on a diet at the time. When the *Wall Street Journal* reporter meets him, Phil is about to go on vacation in Cancun, and his goal is to whittle off a few pounds to improve his beach profile. Phil and his friends meet for dinner at McDonald's, a favorite haunt. Phil orders a double quarter-pound burger with cheese, fries, and a Coke, and also a deep-fried chicken sandwich, a second order of fries, and a chocolate shake. He tells the server to "super-size it all," paying an extra seventy-eight cents for the favor. He leaves the counter with something like 3,200 calories on his tray, and polishes it off in minutes. Leaning back, he sips the dregs of his Coke and shares his weight loss strategy: it's the chicken sandwich. "I'd feel like a fat ass if I ate two burgers," he says.

Here is the object of Phil's affection: two frozen, thawed, and flash-broiled disks of beef topped with two slices of American cheese slathered in a sauce of soybean oil, egg yolks, and high-fructose corn syrup perched on a bun composed of more sweetener, white flour, and oil. It's hard to even imagine concocting something like that at home. But packaged and sold for a couple of bucks, this blur of fat, sugar, and cheap protein screams "massive good deal." Phil is understandably value-conscious—he wants a lot of food for his money. Super-size and value meals are low on vitamins, minerals, and fiber. What they have plenty of is sugar, fat, and calories—more calories than most people need in a meal or, as in Phil's "diet meal," more than we need in a day. But fast food, like so much restaurant food, is not about human needs. It's about desire.

In his collection of postmodern prose poems, *Letters to Wendy's*, Joe Wenderoth slyly captures the bittersweet allure of fast food's cheap excess. In "Today I Had a Biggie" he writes:

> Usually I just have a small, and refill. Why pay more? But today I needed a Biggie inside me. Some days, I guess, are like that. Only a Biggie will do. You wake up and you know: today I will get a Biggie and I will put it inside me and I will feel better. One time I saw a guy with three Biggies at once. One wonders not about him but about what it is that holds us back.

It is the manifest destiny of fast food purveyors to position their wares within arm's length of every human being who can afford to buy. McDonald's trademark golden arches currently stand outside more than twenty-nine thousand franchises in 120 countries around the globe. Roughly two thousand new outlets sprout each year, stretching from Cape Town to Calgary, spreading the American way of eating around the globe. McDonald's first Russian outpost, opened in 1990 near Moscow's Pushkin Square, remains today the world's busiest fast food outlet. At the same time, childhood obesity and overweight in Russia, once hardly a problem, rose to 16 percent.

Fast food is not the only culprit. Restaurant meals of all kinds generally contain more than twice as many calories as do home-cooked meals: any three-star chef worth his toque knows that butter and cream—not purified vegetable matter—is what generally sells $40 entrees. But for most of us, three-star meals are not the problem. It's the no-star meals that add up, the sort of meals we would never cook for ourselves, but, thanks in part to the "super-size" standards of the fast food industry, we have come to expect when we eat out of the home. A scoop of tuna or egg salad flattened between slices of bread and garnished with a pickle was once a perfectly respectable lunch counter staple. Today that same lunch seems woefully stingy. A generation ago, a Pepsi was seven or eight ounces. Today an average serving of soda is twenty ounces. Automakers have widened the circumference of cup holders to accommodate the twenty-ounce bottles, but not enough to accommodate the "Beast," an eighty-five-ounce refillable cup introduced by the Arco service station chain in 1998. Twenty years ago a fast food burger came as a fairly harmless splat of beef on a bun, and an order of fries was a mere handful of mischief. Today's fast food burgers are objects of wrist-wrenching heft.

Fast food creations have indelibly changed our concept of portion size. Restaurant plates have ballooned from ten-inch disks to platters a full foot in diameter. Heavily promoted appetizers like deep-fried "Bloomin' Onions," "cheese fries," chicken fingers, and nachos constitute 30 percent of restaurant profits, and add as much as three thousand calories and several days' worth of fat to a meal that is already over-rich. Clam shacks, Chinese, Tex-Mex, and Indian restaurants, "family style buffets," and steak and pancake houses boast of huge—even limitless—portions of cheap, easy-to-eat food.

The growth in fast food explains, at least in part, the decline in per capita consumption of simple fruits and vegetables in favor of highly processed ones. For example, our consumption of fresh potatoes has plummeted, but we eat almost four times as much processed "potato product" in the form of chips, French fries, and frozen hash browns than we did thirty years ago. Compared to just twenty years ago, we eat thirteen more pounds per person per year of cooking and spreading oil, in particular hydrogenated vegetable oil in processed foods. We eat nearly triple the amount of cheese—the dramatic increase due to the growing fondness not for runny Camembert or pungent Roquefort, but for "cheese food" and other mild cheese mixtures melted onto burgers, pizza, pasta, and nachos. Pizza Hut not only tops but also stuffs the crusts of its pies with cheese. "Poppers"—deep-fried breaded mozzarella balls—are standard on many "Italian style" menus. Cheese-drenched corn chips are a "Mexican" standard. Cheese has even found its way into sushi, in a boggling perversion of the form known as "spicy tuna rolls." Indeed, cheese has become an imperative, the common denominator in our melting pot cuisine. And bacon, that luscious, nostalgic crunch of smoke-flavored fat, has become such a popular add-on that a couple of years ago the fast food industry was credited with almost single-handedly reversing a collapse in pork prices. Bacon sales have more than doubled in the past decade, due in large part to the introduction of breakfast sandwiches and the promotion of bacon as a garnish on tacos, burgers, and salads by the food-service industry.

The popularity of fast food transcends culture, or, perhaps more accurately, dips below the radar of cultural distinction. McDonald's alone has 30 million customers worldwide in Korea, New Zealand, India, the Philippines, Brazil, and Indonesia. By 2003, the company intends to have six hundred outlets in the People's Republic of China. Harvard anthropologist James Watson points out that in many parts of east Asia the term "fast" has been subverted—the food is delivered fast, but it is consumed slowly, sometimes over a period of hours. Fast food restaurants in China and Japan have become clubhouses for the young, places not to gulp and run, but to hang out with friends of both sexes. Like shopping malls, they are the new agora. To some degree, the food is tailored to local taste and custom—restaurants in India serve mutton Maharaja Macs and vegetable McNuggets, and in some Israeli outlets Big Macs are served without cheese out of respect for the Jewish prohibition against mixing meat and dairy products. But the sandwich, fries, and Coca-Cola fast food formula has become a staple throughout the world, as has the focus on quick, cheap, and easy-to-eat meals.

French cultural historian Claude Fischler summarized what he regards as the universal appeal of fast food: "Without doubt, the planetary success of McDonald's, of fast food in general, and of pizza in particular, points to a certain number of alimentary 'universals.' Fast food is not purely function and the customer does not eat it only for reasons of convenience, price, and time. In fact the repertory of tastes and textures it provides devolves upon a kind of least-common-denominator of preferences. In the softness, the 'give' of the hamburger rolls, in the chopped beef, the sweet sauces and the salt-sweet ketchup, one recaptures infantile experiences, regressions, and transgressions." In Asia, where traditional meals are never taken on the run and are always served with rice, cheap fast food offerings are considered snacks, and Chinese and Korean men in particular report that they find the food "unfilling" and "less than satisfying." Disqualified as a serious meal in these societies, fast food is eaten in addition to meals, making its contribution to the growing obesity problem all the more powerful.

Adam Drewenowski is a psychologist who directs the nutritional sciences program at the University of Washington School of Public Health and Community Medicine. Trim, fastidious, and a self-confessed "foodie," he has, for decades, studied the connection between obesity and what psychologists call the "hedonic impact" of food. He makes the case that value and choice are not absolutes, but a matter of perception. Genuinely flavorful, healthy food is not cheap, and is rarely, if ever, part of the "added value" of super-sized meals. "They're not super sizing lobster or anything like it," he says. "You're getting extra French fries, more soft drink, cheap stuff that is essentially filler. No one is offering you a large salad for the price of a small one. You think you're getting a deal here, but you're really not." Salads are rarely part of these "value meals" and for good reason—fresh vegetables are pricey to buy, prepare, and store. (Salads also take longer to eat, anathema to the fast food world.) Servers are trained to remind customers of the "good deal" they'll get by super-sizing their meal for the very sound reason that super-sizing is not a good deal. A large Coke contains roughly three cents' more syrup than does a medium Coke, a super-size order of fries only a few pennies more in fat and starch than a regular order. The restaurants profit from super-sizing because customers are charged far more than a few cents for the extra food they probably wouldn't want were they to know what they getting. Yet the perception of super-size as bargain is ubiquitous—a triumph of quantity over quality.

When a "value meal" of double cheese and mayonnaise-slathered burger, large fries, and a quart of cola is offered, how many of us see as realistic opting for the much more expensive broiled chicken sandwich and side salad, particularly when we have a family to feed? A double burger or sausage sandwich that costs only a buck fills up a noisy child with little fuss, while a salad costing two dollars or more might provoke resistance. When budgets and schedules are tight, we "choose" to go with what is quickest, cheapest, and easiest.

"People respond to the economics of food choices," Drewenowski says. "Lab scientists aren't able to handle this concept—for them, talking about the price of food is taboo—but it's extremely important. Americans spend about three dollars and seventy-five cents a day on food that they eat at home. They can eat pizza at about a thousand calories a dollar, or Oreo cookies at about twelve hundred calories a dollar. M&M's, at about three thousand calories per dollar, are a huge bargain. Spinach is about thirty calories a dollar, not a bargain. And don't even think about lettuce or cucumbers or tomatoes or, heaven forbid, strawberries—by comparison, those foods are a rip-off! Nutrition educators tell us how to eat, but they don't give us the money to change our behavior. Given that people don't really want to spend more on food, I don't see that they have any choice here . . . the choice is really made for them. Rats in a lab have a choice. But humans are constrained by costs."

While adherents claim that there are healthy choices to be made among fast food offerings, these choices are difficult to ferret out. Many of the presumably healthy offerings are not. The Taco Bell Taco Salad with salsa has 850 calories and 52 grams of fat, the fried Burger King Chicken sandwich 710 calories and 43 grams of fat. Even the virtuous-sounding fruit and yogurt parfait at McDonald's contains 370 calories, the great bulk of them from sugar. And by excluding salad and most other healthy options from their value meal menus, purveyors of fast food can claim concern for the public health without backing it up with realistic choices. Were a salad to be included in these meals, it is certain that more of us—perhaps many more of us—would eat it, and perhaps eat a bit less of the less healthy selections.

In Karelia, a region of Finland where obesity and heart disease are serious concerns, the government subsidized free salads at restaurants and workplace cafeterias. This helped double vegetable consumption, and contributed to a 73 percent decline in cardiovascular disease over a period of several years. By making fresh vegetables so widely available and encouraging people to develop a taste for them, the government contributed to an overall acceptance of vegetables and an increase in their consumption. Preference in food—like preference in fashion or film—is not inborn. It is cultivated.

Gary Beauchamp is deeply interested in matters of taste. A psychologist, he is president and director of the Monell Chemical Senses Center, a research institute loosely associated with the University of Pennsylvania, located squarely across the Schuylkill River from downtown Philadelphia. The otherwise anonymous complex of labs and offices is distinguished by an enormous gilded sculpture of a human nose and mouth lurking just in front of its entrance. Inside, the main order of business is made obvious by the pervasive odors wafting through the halls. On the day I visit, it smells of various things, the overall olfactory impact being that of a large cage in which a family of untidy hamsters is baking chocolate chip cookies.

Beauchamp tells me that we are born with only a handful of taste prejudices—among these, an aversion to bitter and a preference for sweet and salt. (Scientists quibble over whether the taste for monosodium glutamate—MSG—is also inborn, but that's a wrinkle.) Most humans, particularly young children, are extremely suspicious of novel tastes. Parents are painfully aware of this and the clever ones learn to disguise new foods in old—for example, conceal their toddler's strained peas in a dollop of applesauce. After a while, the ratio of old familiar applesauce taste can be slowly diminished until the baby will, in some lucky cases, take her peas neat.

No one is born with a taste for hot, bitter, or sour, or, for that matter, a preference for unblended Highland Scotch or Cuban cigars. Such tastes develop with exposure and social pressure. A dramatic illustration of this is the preference in certain cultures for hot peppers. Chili peppers contain capsaicin, an irritant that elicits a searing sensation in the mouth. Small children wisely avoid eating anything that causes them pain. But at age five or six, Mexican and Indian children begin to cultivate a taste for hot chilies, perhaps as a rite of passage. Asian children eat sour plums, African children bitter greens. Developing a taste for these traditional foods is encouraged through repeated and consistent exposure. But when children dictate family food choices, as is increasingly the case in the United States, entire households are immersed in a miasma of one-dimensional sweet taste that reinforces and entrains juvenile preferences. For this reason, marketing soft, sweet, and salty foods is good business, particularly when those foods are aimed at young children, who are still in the process of not only shaping their own tastes, but also influencing the tastes of their parents and siblings.

In the matter of the obesity epidemic, the sugar industry argues that sugar per se is not a factor, and on some levels it is right. At about eighteen calories a teaspoon, sugar is not particularly fattening; no one gets fat eating LifeSavers. The problem is that we are well beyond the teaspoon and LifeSavers stage. A medium chocolate malt shake at Dairy Queen packs an astonishing 131 grams—nearly five ounces—of sugar. Coffee drinks are served with "shots" of syrup, ice cream is embedded with chunks of candy, yogurt is marketed with accompanying sweet toppings. Consumption of sugar in the United States is currently at a record level: thirty-four teaspoons daily per person by the most recent USDA estimates, a nearly 30 percent increase from only fifteen years ago. Annual consumption of caloric sweeteners—which includes not only beet and cane sugar but also high-fructose corn syrup, dextrose, and glucose—has jumped by thirty-two pounds per person since 1970, thanks in part to the enormous and still growing popularity of soft drinks. Americans drink five times as much soda today as they did in the 1950s, making soda the single major source of added sugar in the American diet. "Liquid candy," as the Center for Science in the Public Interest calls it, contains about ten teaspoons of sweetner in a twelve-ounce can, just two teaspoons short of what the government recommends we consume in an entire day. From 1970 to 1997, soda consumption soared from twenty-one to fifty-six gallons per person per year, and today two out of three adolescents drink at least one can a day. As a result, most American children and young adults vastly exceed the recommended limit of added sweetener.

The sweetener lobbies argue strenuously that sugar and corn syrup consumption has no relationship to obesity, and have funded several studies to support that contention. But these findings are made suspect by less compromised findings such as those of Harvard endocrinologist David Ludwig, who concluded in a study published in the British medical journal *The Lancet* that at least for overweight children, soft drink consumption directly predicts weight gain.

Ludwig, director of the obesity program at Children's Hospital in Boston, says that children arriving at his clinic consume as many as a thousand calories a day in soft drinks. When I visit his clinic, a thirteen-year-old boy is being weighed by a nurse—he is tall and broad and weighs more than three hundred pounds. He is also diabetic. The boy won't talk about himself, but his mother tells me that he eats only one meal a day—a healthy one, she says, in front of the television set—washed down with soda, maybe a couple of bottles of soda, big bottles. She describes this as "what any boy his age would drink, you can't expect him to settle for water." She says that he probably buys another bottle of soda on the way to school for breakfast, and maybe another at school, after lunch.

Soft drink companies are notorious for their aggressive marketing tactics, and in particular for their pacts with school districts to share profits from sales in exchange for exclusive "pouring rights." Lifetime Learning Systems, a Stamford, Connecticut firm that specializes in product placement in schools, writes in its promotional material that "School is . . . the ideal time to influence attitudes, build long-term loyalties, introduce new products, test-market, promote sampling and trial usage and—above all—to generate immediate sales." The soft drink industry clearly agrees. Testifying to the importance of school sales in 1994, a spokesman for Coca-Cola explained, "Our strategy is ubiquity. We want to put soft drinks within arm's reach of desire . . . schools are one channel we want to make them available in." Historically, schools have eagerly colluded in this effort, and have joined soft drink companies in pressuring Congress to allow school sales of soda and other snack foods over the continued objections of the USDA.

In 1998 the North Syracuse (New York) Central School District signed a ten-year agreement with Coca-Cola that required all ten of the district's schools and preschool programs to use Coca-Cola products exclusively in all vending machines, athletic contests, booster club activities, and school-sponsored community events. In exchange, Coca-Cola guaranteed a payment of $1.53 million and additional commissions on purchases exceeding the guaranteed minimum in sales. This sort of deal is commonplace—roughly one-sixth of school districts across the country swap valuable vending machine and cafeteria shelf space for computers, team uniforms, and other luxuries—and even necessities—that otherwise would go lacking.

The complaints against this practice from the USDA, the Centers for Disease Control and Prevention, and elsewhere have been loud and long, empowering school districts in nearly a dozen states to avoid what they consider to be a Faustian bargain. In California, legislators are working toward a law that would prohibit anything but milk, water, or juice from being sold to elementary school students, and curb the availability of soft drinks for older students. In Hawaii there is talk of ridding vending machines of soda altogether, and in North Carolina, lawmakers are calling for an end to school contracts with soft drink makers. Schools are among the staunchest opponents of such restrictions, claiming that the soda money is necessary for the smooth functioning of underfinanced programs. When the USDA attempted to ban candy and soda sales in schools more than two decades ago, some school districts joined forces with the National Soft Drink Association to thwart the effort in federal appeals court. In Senate testimony nearly a decade later, industry representatives argued, "There is no need for 'Big Brother' in the form of government injecting itself into decisions when it comes to refreshment choices." But again, just who or what constitutes Big Brother is a matter of perspective. At the university where I teach, such a deal has been made, and snack bars, cafeterias, and vending machines stock only one brand of soft drink. Many employees and students do not favor this brand, and it seems to them that rather than allowing freedom of choice, such arrangements restrict choice. And by striking these deals, school systems have helped shove non-soda beverages to the sidelines.

A generation ago we drank twice as much milk as soda. Today that ratio is reversed, and it is a staggering fact that in the last decade, soda eclipsed coffee and tap water *combined* as the American beverage of choice. Despite the proliferation of coffee bars, coffee drinking in the United States plummeted from thirty-nine to nineteen gallons per capita between 1969 and 1998, leading coffee makers to position coffee as a cold, sweet drink, much like soda. Concoctions like Starbucks White Chocolate Mocha, with 600 calories, and Dunkin' Donuts Vanilla Bean Coolatta, with 400 calories and 80 grams of sugar, are favored for the very reason that they entirely lack the bracing bitter jolt of coffee.

Preference for sweet has become indelibly etched in the Western palate, and the incentive to "outgrow" this preference has weakened. This is increasingly true in non-Western cultures as well. In the Japanese language the word *amai,* or sweet, connotes "cloying." There isn't another more positive word for sweet in Japanese, and traditionally a fondness for sweets is looked down upon as childish. Japanese children eat far fewer sweetened desserts than do Western children, and until quite recently consumed less ketchup, jam, and other sweet condiments. But this is changing as the Japanese and other Asian cultures are increasingly exposed to Western foods, which rely on sweet taste rather than the complexity of flavors offered by traditional Asian cuisine. Soft drinks are threatening to overtake the tea culture in India and Indonesia, and seem to have already done so in the Philippines, where per capita Coca-Cola consumption is the highest in the world, and where obesity rates have risen from negligible to more than 10 percent in just the past decade. In Japan, where obesity is also on the rise, the popularity of fast food restaurants has loosened the prohibition against *rappa nomi*—that is, guzzling cola straight from a can or bottle as though blowing on a trumpet (*rappa*)—at least for the young.

Peter Havel, an associate research professor at the University of California, Davis, says that soda in particular may contribute to weight gain because of its high fructose content. Havel and others have shown that insulin secreted by the pancreas stimulates leptin production in fat cells—so when insulin secretion is stimulated by consumption of carbohydrates composed of glucose, a natural satiety mechanism kicks in to turn down appetite. But the high fructose content of soft drinks, sweetened fruit beverages, and some other processed foods results in less insulin secretion, and therefore reduced leptin levels, compared to consuming carbohydrates composed of glucose. Fructose-laced foods and beverages, Havel says, essentially muffle the leptin response. This may help explain why, for most people, the consumption of sweetened soft drinks does not dim appetite. "Given that high-fructose consumption in the United States has doubled in the last three decades, there's a good chance that this could be part of our problem," he says. And since table sugar—sucrose—is also half fructose, the incredible increase in overall sugar consumption has almost certainly added to the problem.

Scientists have long debated whether particular components of food have psychoactive functions, and if so, whether these changes can influence the desire for particular foods. Underlying this question is whether certain food components, like certain drugs, are addictive. Sarah Leibowitz thinks the answer is yes. A neuroscientist at Rockefeller University, Leibowitz has found in animal tests that repeated and frequent exposure to fatty foods reconfigures the brain to crave still more fat. She says that varying the amount of fat and other nutrients in the diet alters brain chemistry in rats by changing the expression of genes, and that this change in gene expression redirects dietary preferences. At least two-thirds of Americans eat more than the recommended percentage of their calories in fat, and with good reason. Many aromatic compounds that give foods their distinctive flavor are fat-soluble, and fat is the vehicle that delivers these flavors to our taste buds. Fat is cheap, and can make palatable even dubious foods—like unctuously sweet Dairy Queen milkshakes and heat lamp-shriveled burgers. (This is why most fast food sandwiches come with a thick layer of fatty "special sauce.") Leibowitz and others have found in animals that consuming a diet that is more than 30 percent fat elevates the desire for fat and carbohydrates on a physiological level. If the same holds true for humans, our high-fat diet almost certainly contributes to a vicious cycle of overeating.

John Blundell, research chair in psychobiology and chair of the department of psychology at the University of Leeds, has studied mechanisms of appetite control for more than twenty-five years. While not prepared to agree that eating fat can retool the brain, he does say that fat delivers such a dense package of calories that it overrides the body's system to sense satiety. "Exposure to fat induces a liking for fat, and people can eat a huge amount of the stuff before inhibitory signaling systems come into effect," he says. "We are able to eat 150 grams [1,350 calories] of fat in one sitting. Were the inhibitory systems working as you'd expect, there is no way we could eat that much. Eating fat seems to blunt the inhibitory effects of leptin and other signals. It simply messes up the system. Also, as people age and become sluggish, the feedback mechanisms in the brain slacken, and appetite control weakens. But we're not dealing with a biological inevitability here; some people are better able to tolerate a high-fat diet than are others."

Peter Havel takes this one step further. Fat, he says, has even less impact on insulin secretion than does fructose. "When you eat complex carbohydrates you stimulate the body's production of insulin and leptin," he says. "A high-fat diet is far less effective at stimulating insulin and leptin. In the brain, this is interpreted as a relative deficiency of leptin, and over a long period of time, this can lead to overeating."

Complicating matters is that the hedonic effect of food—its pleasurable aspects—can overpower all but the strongest satiation signals. Overweight rats raised on a high-fat "cafeteria" diet similar to the one most Americans confront do not eat less when injected with leptin. But when those fat rats are switched to a diet of boring rat chow, they eat less than do normal rats raised on rat chow. The fat rats have become so acclimated to their luxury diet that when the fat is taken away, their appetite mutes.

One can certainly consume as many or more calories on a low-fat diet as on a high-fat one, but few people do, especially overweight people. As accustomed as we have become to fat, a low-fat diet just does not interest us, and like the fat rats, we tend to eat less of it. That is why very low-fat diets work so well, at least temporarily. Whether dieters eat a high-carbohydrate low-fat diet or a high-protein low-fat "Zone" regimen doesn't really matter—once the fat is removed or sharply limited, the food becomes less palatable, we eat less of it, and the pounds drop off. But few of us are able to resist for long the varied, highly palatable selections made available to us. Like Marie Antoinette, we've tasted cake—and Double Stuf Oreos—and can't be expected to live life on the human equivalent of rat chow. Given the choice, most rats wouldn't either.

This hasn't kept food makers from using low fat as a selling point. Every year for a decade one thousand new low-fat and reduced-fat products swamped the market, yet fat consumption has decreased—if at all—only slightly. We seem to have cut back the *relative* amount of fat in our diet from about 40 percent of total calories thirty years ago to about 34 percent of total calories. But while the percentage of fat is lower, the absolute amount of fat we eat has held steady. We have managed to lower the *percentage* of fat in our diet largely by raising the amount we eat of everything else.

Thanks to the overabundance of cheap, readily available food, Americans today eat 2,002 calories a day compared to 1,854 calories a day in the late 1970s. Theoretically, those 148 additional calories add up to a frightening fifteen pounds of body fat *per year.* Fortunately this is not precisely the case. First, Americans on average have become taller, and therefore require more calories. Secondly, overeating activates the sympathetic nervous system, and boosts thermogenesis, the generation of heat, which burns calories. The process of chewing, swallowing, and digesting also burns calories. So while these additional calories are certainly contributing to the obesity epidemic, they are not contributing quite as much as might be feared.

Humans, of course, show considerable individual variation in response to overeating. Generally, scientists agree that metabolic rate is a very poor predictor of obesity, but what is almost certainly a potent factor is non-exercise activity thermogenesis, or NEAT. NEAT is what most of us think of as nervous energy—the fidgeting, restless pacing, maintenance of posture, and other subliminal activities of daily life. For reasons not yet understood, some people sharply increase these unconscious exertions in response to overeating. James Levine, an endocrinologist at the Mayo Clinic in Rochester, Minnesota, has for years been interested in factors contributing to weight gain. In 1999, he and two colleagues conducted a study in which sixteen volunteers (including Levine himself) were overfed a thousand calories a day for two months. On average, the volunteers gained ten pounds. But as is so often the case in obesity studies, the average was not particularly revealing. One subject gained only two pounds, while another gained sixteen pounds. Levine sorted through a range of factors and concluded that differences in NEAT levels accounted for a *tenfold* difference in fat storage among the volunteers. Other researchers have found similar effects, leading experts to conclude that NEAT is perhaps the most important factor in determining individual differences in response to overeating, at least over the short term.

Just as fidgeters are, to a degree, protected from weight gain, so are those who more readily burn fat rather than carbohydrates as fuel. This is true in part because storing carbohydrate as body fat requires 15 percent to 20 percent of calories eaten, while storing dietary fat as body fat requires only 3 percent of calories consumed. Therefore, an apple, which has approximately one hundred calories of carbohydrate, requires fifteen to twenty calories to be converted and stored as body fat. A pat of butter, also one hundred calories, requires only three. Hence, in the true sense, fat is more fattening than carbohydrate, not only because it carries more calories per gram, but also because the calories it carries "cost" the body so little to store.

Burning fat first means forcing the body to store "expensive" carbohydrates as fuel. Steven Smith, an endocrinologist and researcher at the Pennington Biomedical Research Center in Baton Rouge, Louisiana, has found that fat-burning capability can be manipulated. When I stop by his office for a chat, Smith runs us both up several flights of stairs to show me the metabolic chambers he uses to determine in what proportion research subjects burn protein, fat, and carbohydrate. These chambers are in principle not all that different from the one used by Wilbur Olin Atwater to measure metabolism more than a hundred years ago, though a lot more comfortable and a good deal more accurate. Using these chambers, Smith has found that some lucky folks are born fat burners: their bodies almost always seem to burn off dietary fat before they move on to carbohydrates. Most fat burners are exercisers, the sort of people who, like him, bolt stairs two steps at a time. This, he says, is proof that it's not just a high-fat diet, but the interaction between these diets and our sedentary lifestyle, that promote obesity.

"Some of us are physically fit genetically, we automatically burn fat easily," he says. "But the rest of us can increase our fat-burning capacity by just getting up off the sofa."

Getting off the sofa—it sounds so simple. After leaving Smith's office, I tug on gym clothes and sneakers and head to the Pennington Biomedical Research Center fitness center. The place is magisterial, square foot upon square foot of gleaming, chrome-crusted machinery. I'm contemplating the Stairmaster when an attendant appears, dressed in stiff white polo shirt and shorts. She carries a clipboard. "I'm afraid this equipment is for research purposes only," she tells me. The irony escapes neither one of us, but I don't argue. I decide to jog, and drive back to my hotel, walk out to the street and wait for the traffic to thin. It doesn't. No street lights, no sidewalk, just streaking headlights, not necessarily lethal, just a bit too risky. I check my watch, consider my options. It's nearly six o'clock, and Baton Rouge is a fine food town. I pad to my rental car, rev the engine, and steer toward dinner.

ELEVEN

THE RIGHT CHOICE

Fast, easy and simple are very appealing. They appeal to our lethargy, lassitude and laziness. Hard, slow and complicated on the other hand seem to be associated with most of those things we most cherish in our civilization—great literature, great art, great music, a functioning democracy.

—Benjamin R. Barber

Nancy Wright wasn't expecting miracles when she had her stomach stapled. For a while though, she thought she'd been handed one: she lost 125 pounds in eighteen months. She bought new clothes, got a new job, and started to feel pretty good about herself. Then she hit upon the trick of melting crackers slowly on her tongue, and found that she could suck down quite a few without upsetting her newly abbreviated stomach. Ten pounds have already crept back, and she is starting to wonder where this will end. She has cause for concern: her daughter lost ninety pounds after gastric bypass surgery, and over the past year has gained every one back.

Obesity resists easy remedy for good reason—the human body evolved biochemical redundancies to protect at all costs the instinct to nourish itself. Rudy Leibel, Jeffrey Friedman, Steve O'Rahilly, and hundreds of other scientists have helped illuminate the neurochemical pathways of appetite. We now know that the network in the brain that regulates our intake of food is inextricably linked to our capacity to reproduce. It is, in a word, fundamental. It is part of who we are. But is not immutable. Recent advances in molecular biology and physiology have made possible if not miracles, then truly wonderful things. Leptin therapy has cured the obesity and saved the lives of the Punjabi cousins, and it will cure others who harbor the same genetic defect. Science is almost certain to design cures for victims whose obesity stems from other single gene anomalies, and to develop other weight loss therapies, in particular safer and more efficacious drugs. MLN4760, the code name for a new drug that interferes with an enzyme that seems to encourage fat storage in mice, is, as I write, undergoing human trials in Britain. MLN4760 could be unsafe, or it may not work, but excitement about the substance is justified, for it is the first drug built back from a specific gene—the first product of genomic technology—to be tested on humans. There are other such drugs in the pipeline. One or another of these preparations may one day mitigate the symptoms of obesity, and the diseases linked to it. But this is not at all certain. Nor is any weight loss drug likely to permanently alter human physiology or brain chemistry. To be effective against obesity, drug treatment must be continuous, and administered under a watchful eye. Given the scope of the obesity pandemic, this is not a particularly hopeful—or realistic—prospect.

Obesity represents a triumph of instinct over reason, and as such it embarrasses us. We prefer to think of ourselves as rational beings, in firm control of our destinies or, at the very least, of our bodies. But the diciphering of the genetic underpinnings to weight regulation has ascertained that appetite is to some degree biological, and that our drive to eat can sometimes eclipse reason. There is no longer doubt that some of us more than others are inclined toward overeating, and as a result, toward fatness.

But in the case of body weight regulation, biology is not destiny. A genetically determined "set point" has not been found in the human brain. What we are more likely to have is a settling point, determined by the sum of our genes interacting with a particular environment. Studies by David Barker and others of infants conceived and born in famine give evidence that prenatal and early life play a role in this calibration. Observations of populations in transition, such as in Kosrae, offer striking evidence of the power of environmental forces to recalibrate coordinates set by genetic and prenatal factors, to raise the bar on the settling point. As the world becomes increasingly obesegenic, this bar will continue to rise, and more of us will become overweight and obese. Some of us will do so sooner and with more ease than will others, but we are all vulnerable. After all, the obesity pandemic is less than two decades old. Human genetics hasn't changed in that time. But the environment acting on those genes has changed profoundly.

Humans are wired to roam; we tend to "allow" ourselves from sixty to about ninety minutes a day for "business" travel. This "law of constant travel times" has deep roots. Some speculate that as territorial animals we are somehow compelled to patrol our chosen territory, but prefer to stay within an hour's journey from home, just in case. Whatever the reason, many of us seem to cling to this inclination. In preindustrial days, we could walk about three miles in the appointed hour. Exertion was considered cheap then, and we thought nothing of squandering it on a walk to visit a friend. Today an hour's travel distance has multiplied many times over—but very little of the journey requires human effort. We patrol our territory and do most of our work and much of our play by machine. Few of us would walk more than a mile to chat with a friend, for it is so much easier to phone, fax, or e-mail. Expending human energy seems to require a commitment of money, time, or both that many people cannot or believe they cannot afford. Even those of us who want to often find it difficult. Everywhere there are obstacles to human exertion. In many parts of the country, sidewalks are a rare curiosity. Shopping centers are concrete islands surrounded by highways too treacherous to access by foot. Workplaces are inaccessible by public transportation. School districts, concerned over safety and litigation, prohibit children from biking to school, and budget cutbacks reduce the number of street crossing guards, making walking unsafe. Libraries, parks, and playgrounds are surrounded by acres of parking lot, making their access by anything but car if not out of the question, then at best unappealing.

Even those willing to risk it are put off by the growing distance we have put between ourselves both from others, and from the places and things we need and desire.

Atlanta, Georgia, offers a chastening illustration. Atlanta was once notable for its arboreal charms–its towering canopy of ancient oaks, poplars, dogwoods, maples, and magnolias. But between 1980 and 2000, an average of 110 acres a *day* of forest and farmland were flattened to make way for thousands of housing developments, strung like afterthoughts along ribbons of highway. Today the city is still leafy in parts, and is blessed with some elegant neighborhoods. But taken as a whole, greater Atlanta is gangly and awkward, splayed out over an area larger than the state of Delaware, 110 miles north to south. Downtown serves mainly as a hub upon which commuters converge from greater and greater distances. Atlanta's traffic delays are the worst in the nation, but other places like Los Angeles, Houston, Phoenix, Dallas, Seattle, Boston, Washington, D.C., Denver, and Fort Lauderdale are hot contenders.

Sprawl cities are common in the United States and increasingly around the world, in Japan, Thailand, Brazil, India, and China. In many localities, zoning laws have made it easier to build on a city's outer limits where land is cheap. Developers can rely on government funds to eventually subsidize these projects with roads, electricity, and other amenities, paid for by taxpayers. Thanks to this trend, the time Americans spent commuting to work increased by 36 percent between 1983 and 1995. Total vehicle miles grew by 67 percent, most of those highway miles. And foot and bicycle traffic has plunged by 40 percent in twenty-five years. Today the automobile accounts for 85 percent of miles traveled, and most of the remaining miles are covered by airplane or train. In a society where so few of us outside of inner cities walk or bike, it seems increasingly eccentric to do so. How many suburban adults are willing to pedal or stroll to the library or supermarket, knowing that friends and neighbors will look on with amusement from the cozy anonymity of their automobiles? We drive to get a gallon of milk, a newspaper, a greeting card, a video. We drive to collect mail from the mailbox at the end of our driveways. We drive our children to school and to friends' homes and to piano lessons and to basketball practice. Those of us too young or too old to drive depend on the indulgence of driving friends and family. Often, though, the car isn't actually getting us anywhere: the time Americans spend stalled in traffic jumped an incredible 236 percent from the mid-1980s to the mid-1990s. In Atlanta that adds up to fifty-three hours a year—a full working week—sitting and waiting for traffic to clear.

Car culture and its tyranny of false efficiency have contributed mightily to the obesity pandemic, but making this connection does not solve the problem. Cars are inextricably tied to who we are, and we aren't about to abandon them. Communities where residents can walk to shops and parks and to each other's homes are charming and desirable, but for want of money or opportunity, most of us do not live in these places. A reasonable alternative is to retrofit existing communities to meet the needs of human beings outside of our automobile armatures, by installing sidewalks, crosswalks, and streetlights. We should resist the demeaning trend of "parent-as-chauffeur" by centralizing school bus stops within walking distance of neighborhoods, and by insisting that schools hire crossing guards (or recruit parent or teenage volunteers). No park, school, or library should be without a bike rack. Roads should be made safe for our children—and for us—if not with bike and walking lanes, then with firmly enforced regulations that give pedestrians equal consideration to that given cars. Over the longer term, zoning laws and tax incentives can be adjusted to encourage the development of living, working, and commercial spaces within range of public transportation.

In his 1932 essay "In Praise of Idleness," philosopher Bertrand Russell proposed a four-hour workday to ensure that "there will be happiness and joy of life, instead of frayed nerves, weariness, and dyspepsia. . . . Since men will not be tired in their spare time, they will not demand only such amusements as are passive and vapid." The average work week in Russell's time was fifty to sixty hours. Today it is closer to forty. Yet our nerves are as frayed—perhaps more frayed—than ever. And our amusements—largely television—are more passive and vapid than even the prescient Russell foretold. In an economy where productivity is so often ephemeral, success is measured as much in the way we spend our time as in the objects we acquire. Understandably, many of us perceive ourselves to be time-pressed, which is to say, time-poor. We feel we have little time and because of this perception, we are prone to adopt what Russell coined a "cult of efficiency," the belief that "everything ought to be done for the sake of something else, and never for its own sake." We are persuaded our time is too short to spend shopping for dinner or preparing it—though on consideration many of us might actually enjoy these simple pleasures. We are persuaded that spending eleven minutes refueling in a fast food restaurant is "fun" while sitting at table for a family dinner is an "obligation." We are persuaded that walking more than a few blocks is a dreary chore, while spending hours fixated by a cathode ray tube is "relaxing." And we are persuaded of all this by a trillion-dollar leisure industry. The image makers who produce and market the three or more hours of television we watch every day have convinced us that we do not have the time or the energy to do anything but sit down and watch more of their "can't miss" programming.

Time pressure makes us feel frazzled, but that's just the point. Harvard Medical School endocrinologist Jeffery Flier has found evidence that frazzled people make especially good customers. He recently uncovered a connection between an elevation of enzyme activity regulating stress hormones—such as cortisol—and overeating and obesity in mice. Endocrinologists at the University of Edinburgh had earlier found an elevated level of cortisol in the fat tissue of obese men. These findings may one day explain why stress has been linked to obesity—whether brought on by a contentious divorce, the first year of college, or a daily commute in snarled traffic. It's a point punctuated by an A. C. Nielsen survey that showed a spike in snack food sales of more than 12 percent in the few weeks following the World Trade Center tragedy. Suddenly America was awash in "comfort food."

National time use studies reveal that Americans gained almost an hour a day of free time between 1965 and 1985. American men now average about forty hours of free time per week, and American women average thirty-nine. Yet the most common explanation offered for the dramatic increase in sales of fast food and restaurant food is that working parents have neither the time nor the energy to shop or cook. It is true that many of us feel time-pressed, but in part this is because we are told to feel that way. Image makers have codified this "no time" concept in hundreds of thousands of commercial messages designed to convince us that planning and preparing our own meals is a luxury we simply cannot afford. The campaign has been wildly successful. From the late 1970s to the late 1990s, the number of meals and snacks eaten at fast food restaurants increased by 200 percent, and at other restaurants 150 percent. It is unclear how much time we actually save by patronizing these facilities. Leaving home to pick up dinner at the drive-through requires getting into the car, driving, waiting in line, picking up the order, and driving home. All this takes plenty of time. Restaurant and ready-to-eat food is more expensive than homemade, so it is worth considering how many extra hours we work to pay for the perceived "convenience" of take-out meals. What we do know is that these meals are far higher in things we say we don't want—fat, sugar, and calories. And what we also know is that were restaurant and take-out meals to weigh in at the same energy and fat density as home-cooked meals, Americans would be eating, on average, two hundred *fewer* calories a day—which adds up to a staggering seventy-three thousand fewer calories a year, enough to support eleven pounds of body fat.

So many of us eat prepared foods more often than we intend, many scientists contend, because we have become habituated to them. Not all of us become habituated, of course, and those of us who do are not affected to the same degree, but there is strong evidence to suggest that calorie-dense food like fast food can entrain a neurological feedback mechanism that contributes to overeating. Charles Billington, a professor of medicine at the Minneapolis VA Medical Center and the former president of the North American Association for the Study of Obesity, is among the many obesity experts who hold to this view. "The leptin discovery is part of a larger story, that of the role of various brain mechanisms in the control of eating behavior," he told me. "As we develop a full understanding of the neuroregulation of appetite, I think the addictive nature of foods will come clear. And I think we will learn that these addictions can develop at various stages in life, in adulthood as well as in childhood. And I think we will learn that they are very, very powerful."

A generation ago, philosopher Herbert Marcuse warned of the collusion of mass media, advertising, and industry to create false needs in order to integrate individuals into the existing system of production and consumption. Children are the most vulnerable targets. Childhood obesity rates are highest in countries where advertising on children's television programs is least regulated—in Australia, the United States, and England. Americans on average view more than 350,000 television commercials before they reach voting age. These messages are delivered mostly by commercial television, but publicly funded television plays an increasingly important role: to cite but one example, *Teletubbies,* a public television program targeted to the preschool set, is sponsored by McDonald's. Once the television set is on and the channel tuned to children's programming, there is simply no escaping the ads, 80 percent of them for calorie-dense treats. In 1995 the American Academy of Pediatrics condemned advertising to children under the age of eight as deceptive and exploitative. Since then researchers have found that older children may, on some levels, be even more impressionable. With what science has taught us about the habituation of eating preferences, of the near certainty of obese adolescents becoming obese adults, and of the links between childhood obesity and adult-onset diabetes, heart disease, and other health effects, it is time to take stock. Just as there are laws against defacing public property with graffiti, it seems reasonable to support regulations or at least standards to curb the defacing of public airwaves with messages that fuel a growing health crisis. Sweden and Norway maintain a virtual ban on advertising to children, and have consistently low levels of childhood obesity. Ireland, Belgium, Italy, and Denmark pose restrictions on children's advertising, and are pressing the other states of the European Union to do the same. The United States and other countries in the developed and developing world can afford to do no less.

In the United States, as in all civilized nations, we do not permit market forces free rein. We protect our children with child labor laws, the handicapped with antidiscrimination laws, the poor and the infirm with public assistance. We do not allow minors to buy alcohol, cigarettes, guns, or pornography. We do not allow industry to dump toxic waste into the public water supply. We impose these protections because history has shown us that this is the only way to ensure them. Free-market capitalism is wonderful for many things, but public health is not among them.

One need not look far back through history to find evidence for this: it is hard to imagine a surer prescription for ill health than urban life in the nineteenth century. Rapid industrialization led to an unprecedented concentration of people in cities throughout Europe and the United States, overwhelming existing waste collection systems, clean water supplies, and housing. Life in the tenements was brutal and short: tuberculosis, scarlet fever, smallpox, diphtheria, typhoid, and cholera took a horrifying toll. But the appalling conditions of filth and crowding that encouraged and intensified the spread of infectious disease were not the first target of public action and concern. There were rationalizations, questions of culpability, accusations that the victims—particularly the poor—had brought ill health upon themselves. There was moralistic rhetoric, talk of disease as physical manifestation of sin. If cleanliness was next to godliness, it was argued, surely the "sickly classes" had strayed awfully far from God. Could germs really be blamed for all this misery? Deep down at the nub of this soul-chilling rhetoric was the question of money—who would pay to drain the sewers and clean up the filth? Weren't these responsibilities for the individual, or the family? Why should society bear the costs? Would it not be more rational to let the weak die to make room for a sturdier breed?

Not until governments acknowledged the threat of disease to the social and political fabric—that is, to the entrenched interests of the powerful—was action taken, and then only grudgingly and under constant pressure from progressive reformers. It took generations for public health to become integrated into a professional civil service, and still more time for health and safety regulations to be set and enforced. But the civil service was set, the regulations put in place, and infectious disease was on the decline in much of the Western world even before the widespread application of antibiotics and other curatives.

Microbes are the proximal cause of infectious disease, but it was unfettered urbanization and industrialization that engendered the epidemics of the nineteenth century. Overeating and inactivity are the proximal causes of obesity, but unfettered consumerism drives the obesity pandemic of the twenty-first century. Despite well-intentioned government nutrition recommendations, most of us continue to eat more than our bodies can healthfully tolerate. This is thanks, at least in part, to the gut-busting efforts of the food and leisure industries and to a political system that allows a small circle of powerful interests to direct the very policies designed to regulate them.

The elucidation of the molecular foundation of obesity revealed a vicious cycle: the human appetite loses its ability to regulate in an environment offering so few opportunities to exercise and such an abundance of calorie-dense foods. The less we move and the more rich food we eat, the more difficult it is to self-regulate. Unless this cycle is broken, the casualties will continue to mount. Admonitions to "get off the sofa" and "eat a balanced diet" are not enough. Given the scope of the problem, scolding of this sort is of no more use than a Band-Aid applied to a hemorrhaging artery. Twenty-seven percent of Americans are already obese, double the level of just twenty years ago. Without intervention, virtually all Americans will be overweight by 2030, and half will be obese. Other industrialized nations are not far behind. A recent study concluded that obesity has roughly the same impact on health as does twenty years of aging. The cost of this in both lives and in quality of life is not one we should be willing to pay. Turning the tide of this epidemic will require more than rhetoric. It will require rethinking the way we live.

Obesity is the number two public health risk factor in the United States, responsible for the loss of 300,000 lives a year. The number one risk factor is tobacco, at an estimated 400,000 lives, but it is all but certain that obesity will overtake tobacco within the next decade or two. And in many respects, public attempts to tackle obesity mirror the early antismoking efforts. For decades we relied on the tobacco industry to tell us the truth about its products, or, at the very least, on our government leaders to tell us the truth. They didn't. Among Georgia Governor Jimmy Carter's more memorable—and regrettable—1976 presidential campaign promises was, "We're going to do all we can to make tobacco even more safe than it is now!" It may not have occurred to him—and certainly not to most of us—that Big Tobacco would be so cynical. We did not want to believe that the tobacco industry would deliberately engineer products to be addictive, or that it would mindfully prey on children. Big Tobacco was doing that and worse, of course, but it did all in its formidable power to call into question the motives of advocates and agencies that threatened to expose it. So we were fooled, and fooled again.

This is precisely what Big Food is doing today. Like Big Tobacco, it is a cunning manipulator of public opinion. Like Big Tobacco, it characterizes critics as a conspiracy "of food cops, health care enforcers, vegetarian activists and meddling bureaucrats" intent on using "junk science" to build a socialist "nanny state." Like Big Tobacco, it makes us believe that our freedom of choice depends directly on its freedom to garner profits. And like Big Tobacco, it promises to make our lives less frantic while constantly reminding us of how frantic we are. Media messages aim to convince us that we are all cogs on a spinning wheel, too harried to think for ourselves. Many of us sense this manipulation, but nonetheless allow ourselves to be seduced, or coerced into taking what we are told is the easy path. Despite our best intentions, we abdicate control to interests that profit by our feeling powerless.

Historian Richard Kluger wrote in *The New York Times* in 1996, "Even in an era when deregulation is all the cry, many in Congress have begun to grasp that smoking may be an issue beyond partisan politics, and that cigarettes are a product that will never be tamed without government intervention." We no longer rely on the tobacco industry to speak honestly, and as a nation we no longer rationalize our cigarette habit. We actively fight it with public health measures like tobacco taxes, effective government-funded antismoking campaigns, and regulations that almost overnight turned cigarettes from a commodity into a controlled substance. We have not won the war against tobacco, but we have made enormous strides. Tobacco in this country is on the run.

Obesity prevention, too, has been an explicit goal of U.S. public health policy since 1980. But while intentions are good, progress is slight. The halfhearted promotion of vaguely construed "healthful behaviors" and "responsible eating" through educational programs offers evidence that our national priorities lie elsewhere. The 5 A Day campaign for Better Health Program, sponsored jointly by the National Institutes of Health, the National Cancer Institute, and the industry-funded Produce for Better Health Foundation, promotes consumption of fruits and vegetables. The amount allocated to the outreach and communications component of this effort—the part devoted to getting the message out to consumers—is something less than $2 million a year. A $1.5 million National Cholesterol Education Campaign and a $5 million grant for obesity education and prevention to the Centers for Disease Control and Prevention comprise the government's largest contributions. That is a total of less than $9 million. In contrast, the budget for *the initial phase* of a new campaign to sell a single brand of candy bar—the wildly popular Milky Way—is $25 million. Burger King spends an estimated *half billion* dollars on promotional efforts each year. Can there be any question as to which of these "campaigns" is the more effective? We are not losing the war on weight. We have not begun to fight.

Health advocates, and some legislators, have proposed the levying of a "fat tax" to help contain the chronic overconsumption of high-fat foods. Sales of soft drinks, candy, and other snack foods are already taxed to raise revenue in at least eighteen states, and the idea is to extend the practice. This will almost certainly reduce consumption—just as tobacco taxes reduced smoking. And it will generate revenue that could be put to good use backing nutrition awareness programs. As public policy, however, it is problematic. Taxing tobacco in all its manifestations is a relatively simple matter. But there are millions of foods and food products, and distinguishing among them would be terribly difficult, and politically divisive. Donuts and bologna are far higher in fat than are bagels or lean roast beef, but they are also cheaper. Penalizing people for personal taste or for being price-sensitive is likely to result in a backlash.

There is a surer path.

In the past decade, fast food outlets have *increased* substantially the sugar and fat load of their offerings. Were more of us to know this, perhaps consumer interest in these foods would wane. And were purveyors persuaded to make the nutritional particulars of their wares prominent on packaging and on menus, perhaps they would be moved by the public to modify their recipes, and the public to reconsider the "bargain" of super-sizing.

Schools have been turned into virtual sales and promotion centers for Coke, Pepsi, Pizza Hut, and McDonald's. Administrators argue that budget cuts have necessitated these unholy alliances, but the money gleaned is often spent on technology—televisions and computers—that promotes consumption and reinforces inactivity. School districts in San Francisco, Seattle, and elsewhere are actively fighting this trend with great success, and they are to be applauded for their efforts. But no school should be in the position of having to barter the health of its students. Preaching about the importance of exercise and a balanced diet while actively promoting junk food fuels only cynicism.

The push for enforcing academic standards, while laudable, has in some schools forced a cut in lunch periods to carve out more classroom hours, reducing the lunch break to a fifteen- or twenty-minute rush. Rushed eating is not thoughtful eating. Physical education classes have also been cut back or eliminated. Taken together, these forces undermine well-meant efforts by nutrition educators to promote healthy decisions—or any decisions. Childhood obesity is skyrocketing in particular in the most crowded, time-pressed school districts. School lunch offers an opportunity to reinforce the lessons taught in the classroom and in the home. The prospect of teachers or other adults sitting down with students to eat in a calm, dignified atmosphere may seem utopian, but this is precisely the practice in many private schools, and in many schools in France and the Netherlands, where obesity rates are among the lowest in the industrialized world. If this concept seems farfetched, it is not because it is inherently difficult to execute, but because as a nation we choose not to take our children's health seriously.

The old Burger King slogan, "Aren't you hungry?" was rhetorical—we are always hungry because we have come to confuse hunger with appetite. Once toasted as a desired state, something to be "honed" or "worked up," appetite today is too often treated like an illness, and food like medicine, to be swallowed quickly, often alone and on the sly. For many, meals are no longer discrete events, but a blur of snacks. It is time to take stock of this trend, and reverse it. Wherever feasible, we should encourage the cultivation of appetite and mindful eating. Lunch should be a considered break in the day, not a hurried affair eaten hunched over a desk or, worse yet, in a car. Fresh fruits and vegetables should be subsidized in publicly supported food service venues, their price and quality regulated to make them attractive options, rather than the limp and pricey afterthoughts they have too often become. Private employers might consider taking similar steps—offering fruit, vegetables, and salads at cost to employees—as part of their health benefit plan. And vending machines, some even stocked with unprocessed foods, could be clustered together, outside that tempting "arm's reach of desire."

Public nutrition campaigns should go beyond vague recommendations to exercise and eat a balanced diet. Public service announcements targeted to children and to adults should make explicit the link between inactivity, junk food consumption, and obesity. These messages should be designed by the sharpest minds on Madison Avenue, and broadcast not on Sunday mornings at 6 A.M., but during prime time.

The food industry will lobby against these efforts, of course, claiming that they constitute "legislation of food choices." In this they will be supported by nutrition advisory agencies—notably the American Dietetic Association and the International Food Information Council, and the American Council on Science and Health. Like all advocacy groups, these agencies serve an important function for their constituencies: their mission is to insure that we buy—and presumably eat—as much as possible of whatever their clients manufacture. The obesity pandemic offers tragic evidence of the influence and power of these agencies, and also testimony as to why we can no longer afford to enlist them as arbiters of food policy.

The U.S. Department of Agriculture, too, must be relieved of the impossible task of having to both promote the consumption of American-made food products and advising the public on nutrition. The United States runs a food surplus—we grow and process almost double the calories we need. Food companies, in fierce competition to capture public dollars in a flooded market, lobby Congress and the USDA to allow them free rein in their marketing practices. We should break up this unhealthy alliance by separating the food promotion function of the USDA from the nutrition advising function, either by creating another agency or by transferring the nutrition advising arm to another agency—for example, to the CDC's Center for Chronic Disease Prevention and Health Promotion.

Obesity researcher George Christakis remarked more than twenty years ago on the "lively controversy . . . as to whether obesity represents a social embarrassment or a public health problem." As public health scholar Shiriki Kumanyika has pointed out, "it is the either/or nature of Christakis' observation that carries the insight." We can no longer afford to be embarrassed, or to avert our eyes from the forces underlying this tragic circumstance. Science has taught us that the obesity pandemic is less a matter of individual differences than of societal pressures, and of the power of the institutions that impose them. We can and should resist. We can begin by expressing indignation at the conditions that brought us here—the usurpation of our freedom to shape our own and our children's choices, and the imposition of policies that serve only a vested few. The closer we get to raising a sense of moral outrage around these indignities, the less likely they are to stand, and the more likely it is that the obesity epidemic will be, like so many epidemics before it, a thing of memory.

Notes

The Trillion-Dollar Disease

3 "*affecting 1.1 billion adults and a burgeoning number of children*": See Gary Gardner and Brian Halweil, "Underfed and Overfed: The Global Epidemic of Malnutrition," *Worldwatch Institute*, March 4, 2000.

"*34 percent of adults overweight and an* additional *27 percent obese*": See Ali H. Mokdad, Barbara A. Bowman, Earl S. Ford, Frank Vinicor, James Marks, and Jeffrey P. Koplan, "The Continuing Epidemics of Obesity and Diabetes in the United States," *JAMA* 286 (1999): 1195–1200.

"*A body mass index [BMI] of 30 is considered obese, a BMI of more than 25, overweight*": Although a BMI of 27 or above is occasionally classified as obese, a BMI of 30 or higher is the agreed-upon standard of the International Obesity Task Force, the Centers for Disease Control, and other national and international health agencies.

"*Obesity, he warned, is quickly eclipsing tobacco as the number one threat to public health*": See "The Surgeon General's Call to Action to Prevent and Decrease Overweight and Obesity," at www.surgeongeneral.gov/topics/obesity/.

"*Half the adult populations of Brazil, Chile, Colombia, Peru, Uruguay, Paraguay, England, Finland, and Russia are overweight or obese. The same goes for Bulgaria, Morocco, Mexico, and Saudi Arabia*": The World Health Organization keeps careful statistics on the obesity pandemic, made available both on its website and in various publications. The International Obesity Task Force maintains a website with this information at www.iuns.org/features/obesity/obesity.htm. Individual researchers also helped me compile these statistics, in particular economist Barry Popkin, professor in nutrition at the University of North Carolina at Chapel Hill.

"*In Britain, youth obesity rates soared by 70 percent in a single decade*": See J. J. Reilly, A. R. Dorosty, and P. M. Emmett, "Prevalence of Overweight and Obesity in British Children: Cohort Study," *BMJ* 319 (1999): 1039, which documents a 70 percent increase in obesity rates in children ages 1–3 between 1989 and 1998. See also the commentary by William H. Dietz, "The Obesity Epidemic in Young Children," *BMJ* 322 (2001): 313–314.

"*The trend holds true for Russia and China, Brazil and Australia*": See Youfa Wang, "Cross-national Comparison of Childhood Obesity: The Epidemic and the Relationship Between Obesity and Socioeconomic Status," *Int J Epidemiol* 30 (2001): 1129–1136.

3 "*among these are high blood pressure, high blood cholesterol levels, and gyrating insulin levels*": See David S. Freedman, William Dietz, Sathanur R. Srinivasan, and Gerald S. Berenson, "The Relation of Overweight to Cardiovascular Risk Factors Among Children and Adolescents: The Bogalusa Heart Study," *Pediatrics* 103 (6) (June 1999): 1175–1182.

4 "*tends to be awash in significance*": Susan Sontag, *Illness as Metaphor* (New York: Vintage Books, 1977).

"*Americans spend $33 billion a year*": See Graham Colditz, "Economic Costs of Obesity," *American Journal of Clinical Nutrition* 55 (1992): 503–507.

"*We spend hundreds of millions on weight loss medications that can sicken or even kill us*": See "Many Americans Fed Up with Diet Advice," *Agence France-Presse, New York Times,* January 2, 2001: D8.

Chapter 1
A Weariness of Eating

6 "The Egyptian Book of the Stomach": See G. M. Ebers and L. Stern, *Papyrus Ebers* (Leipzig: Engelmann, 1875). Facsimile and partial translation of the ancient Papyrus Ebers, thought to be the oldest extant medical document and dating to 1552 B.C., as cited in John F. Nunn, *Ancient Egyptian Medicine* (University of Oklahoma Press, 1996). See also www.osirisweb.com/egypt/papyrus.html (Grafs 188–207 of the Ebers Papyrus comprise the "book of the stomach").

9 "*The surgery is reserved for those with one hundred or more pounds to lose*": See "Gastrointestinal Surgery for Severe Obesity," NIH Consensus Development Conference Statement, March 25–27, 1991.

"*On average, patients shed about 60 percent of their excess*": Estimates of weight loss after gastric bypass surgery range as high as 70 percent of excess weight, but the generally agreed-upon figure is closer to 60 percent. See, for example, L. K. Hsu, P. N. Benotti, J. Dwyer, S. B. Roberts, E. Saltzman, S. Shikora, B. J. Rolls, and W. Rand, "Nonsurgical Factors that Influence the Outcome of Bariatric Surgery: A Review," *Psychosom Med* 60 (May/June 1998): 338–346.

10 "*Gastric bypass kills one out of a hundred patients on the operating table, and not everyone recovers from its complications*": The one in a hundred figure is the one most commonly offered by obesity surgeons, and the figure patients are usually told, but the studies to back it up are few and scattered. Reports of patients dying within a year of the surgery are not uncommon, but it's difficult to assess whether these deaths were the result of the surgery per se or of other factors.

"*There are few controlled studies of the procedure, so no one can speak with authority on its degree of danger. Still, the insurance industry classifies it as "high risk*": See B. E. Casey, K. C. Civello Jr., L. F. Martin, J. P. O'Leary, "The Medical Malpractice Risk Associated with Bariatric Surgery," *Obes Surg* 9 (5) (Oct. 1999): 420–425. See also: J. S. Torgerson, and L. Sjostrom, "The Swedish Obese Subjects (SOS) Study—Rationale and Results," *Int J Obes Relat Metab Disord* 25 Suppl 1 (May 2001): S2–4. The Swedish Obese Subjects (SOS) study, an ongoing multicenter study of obesity, includes the closest thing to a comprehensive long-term controlled study of obesity surgery. Two thousand surgically treated patients are being compared with a matched group of two thousand conventionally treated patients over a period of ten years. So far, the first group of patients (369 surgically treated patients compared with 371 controls who did not get surgery) has shown positive affects of surgery, including sustained weight loss of an average of about forty-five pounds over a period of eight years, and a sharp decline in diabetes, though not in hypertension. It is premature to speculate about long-term outcomes from this single, yet incomplete study, but it is worth noting that even with an average forty-five-pound surgically induced weight loss, most of the subjects in this study would still be classified as obese.

11 "*seen as a large bag*": Sherwin B. Nuland, *The Wisdom of the Body* (New York: Alfred A. Knopf, 1997).

12 "*Bariatric surgery, as obesity surgery is called, has a controversial history dating back hundreds of years*": See Hillel Schwartz, *Never Satisfied: A Cultural History of Diets, Fantasies and Fat* (New York: The Free Press, 1986), 178–179; Frederick M. Grazer and Jerome R. Klingbeil, *Body Image: A Surgical Perspective* (St. Louis: The C. V. Mosby Company, 1980), 63–64; Joseph Colt Bloodgood, "Possibilities and Dangers of Beauty Operations and the Danger of Excessive Fat in Surgery and Disease," in Morris Fishbein, ed., *Your Weight and How to Control It* (New York: G.H. Doran Co., 1927), 51–66.

15 "*Two interdental eyelets were placed in each canin*": See S. Rogers et al., "Jaw Wiring in the Treatment of Obesity," *Lancet* (June 11, 1977): 1221–1225.

16 "'*balloonacy' in an article in the journal* Gastroenterology": See John G. Kral, "Gastric Balloons: A Plea for Sanity in the Midst of Balloonacy," *Gastroenterology* 95 (1998): 213–215.

"*Michael D. Gershon, author of an ode to the digestive tract,* The Second Brain": Michael Gershon was kind enough to speak with me about the findings and conclusions detailed in his book, *The Second Brain: A Groundbreaking New Understanding of Nervous Disorders of the Stomach and Intestine* (New York: Harper Perennial Library, 1999). The book offers a intimate look at the neuroscience and neurochemistry of the digestive tract, which as Gershon explains has nerve cells that make it a sort of "second brain" that cooperates and interacts with the brain to allow the body to function smoothly.

18 "*Children have said in surveys that they would prefer playmates with missing legs or eyes to those with too much fat*": See C. S. W. Rand and A. M. C.

Macgregor, "Morbidly Obese Patients' Perceptions of Social Discrimination Before and After Surgery for Obesity," *South Med J* 83 (1990): 1390–1395. Discrimination against obese patients is rampant even in the medical community. See, for example, Josephine Kaminsky and Dominick Gadaleta, "A Study of Discrimination within the Medical Community as Viewed by Obese Patients," *Obesity Surgery* 12 (1) (February 2002): 14–18.

19 "*Many obese adults confess*": See S. A. Richardson, N. Goodman, A. H. Hastorf, and S. M. Dornbusch, "Cultural Uniformity in Reaction to Physical Disabilities," *Am Sociol Rev* 26 (1961): 241–247.

"*She knows about the complications*": See, for example, T. K. Byrne, "Complications of Obesity Surgery," *Surg Clin North Am* 81 (5) (October 2001): 1181–1193, vii–viii.

22 "*Often the pounds creep back*": In a review conducted by L. K. Hsu et al., "Nonsurgical Factors that Influence the Outcome of Biatric Surgery" [*Psychosom Med* 60 (3) (May/June 1998): 338–46] about 30 percent of patients regained weight, beginning at eighteen to twenty-four months post-surgery. There were also reports of increased rates of suicide and alcoholism. This article, offering one of the few comprehensive reviews of obesity surgery available, concluded that more research needs to be done to determine the long-term risks and benefits of the procedure.

Chapter Two
Walking Naked

23 "*Willendorf is but the oldest in a long line of Paleolithic Venus figurines found scattered from southwestern France through Italy, Austria, and Turkey, to the north shore of the Black Sea*": See George Bray, "Historical Framework for the Development of Ideas About Obesity," in George Bray, Claude Bouchard, and W. P. T. James, eds., *Handbook of Obesity* (New York: Marcel Dekker, 1998).

"'*palaeo-porn' cult*": See Lee Roy McDermott, "Self Representation in Upper Palaeolithic Female Figurines," *Current Anthropology* 37 (2) (1996): 227–275.

24 "*closer to a truth we are likely never to know*": See Alexander Marshack, *The Roots of Civilization: The Cognitive Beginnings of Man's First Art, Symbol and Notation* (New York: McGraw-Hill, 1972).

"*not from artistic inspiration, but from real life*": See A. S. Beller, *Fat and Thin: A Natural History of Obesity* (New York: Farrar, Straus and Giroux, 1977).

"*Obesity then, may have its roots in the Stone Age*": See R. J. R. Hautin, "Obesity. Conceptions Actuelle," *These pour le Doctorat en Medicine* (Bordeuax: Imprimerie Bier, 1939), as quoted by George Bray in his chapter, "Historical Framework for the Development of Ideas about Obesity," in *Handbook of Obesity* (op. cit.).

24 "*obesity was otherwise unheard-of in traditional hunter-gatherer societies around the world*": See P. Bjorntorp and B. Brodoff, *Obesity* (New York: J. P. Lippincott, 1992), 320.

"*Agriculture had the startling consequence of making food perennially available, first in the Fertile Crescent of the Middle East, then in China, Egypt, and western Europe*": See Jared Diamond, *Guns, Germs, and Steel: The Fates of Human Societies* (New York: W. W. Norton, 1999).

"*Agriculture also made possible the domestication of animals—and the steady consumption of animal fat*": See S. Garn, "From the Miocene to Olestra: An Historical Perspective on Fat Consumption," *Journal of the American Dietetic Association* 97 (7) Suppl (July 1997): S54–7.

"*an unusually fat man, which increased at length to such a degree that he could take no food which was not introduced into his stomach by artificial means*": See Meir H. Kryger, "Sleep Apnea from the Needles of Dionysius to Continuous Positive Airway Pressure," *Arch Intern Med* 143 (December 1983): 2301–2303, referring to C. Aelianus, *Various History: Book IX* (London: Thomas Dung, 1666), 177.

25 "*he lay like a stone; but when it came to the firm flesh, he felt it and awakened*": See Kryger (1983), ibid.

"*reportedly was granted a similar wish, being smothered in his own fat while lying in bed*": See Kryger (1983), ibid.

"*They should, moreover, eat only once a day and take no baths and sleep on a hard bed and walk naked as long as possible*": See J. Precope, *Hippocrates on Diet and Hygiene* (London: Zeno, 1952).

"*massaging him maximally with diaphoretic injunctions, which the younger doctors customarily call restoratives*": See this translation of a section of Galen's *Hygiene* (De Sanitate Tuenda) in Robert Montraville Green, *A Translation of Galen's Hygiene* (Springfield, Ill.: Charles C. Thomas, 1951).

26 "*lists obesity as a disease in his magnum opus* Kitab al-Qanun": See Roy Porter, *The Greatest Benefit to Mankind: A Medical History of Humanity* (New York: W. W. Norton, 1997). A magisterial history of medicine by a gifted and prolific historian.

"*heavy of limb and scandalously stout; with me they are thin, wasp-waisted, and terrible to the foe*": Aristophanes, *The Eleven Comedies,* Volume 2. *Plutus* (New York: Tutor Publishing Company, 1934). Based on translation originally published by The Athenian Society in London, 1912, translator unknown.

"*somewhat perplexing why, in parts of Asia, overeating has for centuries been viewed as a moral failing*": See A. J. Stunkard, W. R. La Fleur, and T. A. Wadden, "Stigmatization of Obesity in Medieval Times," *International Journal of Obesity* 22 (12) (December 1998): 1141–1144.

26 "*Because she ate all kinds of rich foodstuffs, her body became fat and her flesh too abundant*": Ibid.

27 "*whose god is their belly*": See Joseph F. Delany transcribed by Joseph P. Thomas in *The Catholic Encyclopedia,* Volume VI, Online Edition, 1999.

"*where sauces were seasoned with sulfur and devils stuffed gluttons with toads from stinking rivers*": Schwartz (1986), op. cit., 9.

28 "*The Renaissance and the Reformation brought a somewhat less narrow view of sin*": See Raghavan Iyer, "The Seven Deadly Sins-I The Historical Context," *Hermes Magazine* (November 1985).

"*as did a number of monographs written in English*": See George A. Bray, "Obesity: Historical Development of Scientific and Cultural Ideas," *International Journal of Obesity* 14 (1990): 909–926.

"A Discourse on the Nature, Causes, and Cures of Corpulency": See Malcolm Fleming, "A Discourse on the Nature, Causes, and Cures of Corpulency," printed for L. Davis and C. Reymers, printers to the Royal Society, London, 1757.

"*was activated by the sperm of the father*": See Aristole, "The Generation of Animals," available on-line at 222.hol.gr/greece/anczips/g_animal.zip. Translated by Arthur Platt.

"*built upon Aristotle's theory to suggest that all living things originate from the egg*": See Porter (1997), op. cit., 214–215.

29 "*He also proposed that males and females contributed equally to future generations*": See Robin Marantz Henig, *The Monk in the Garden: The Lost and Found Genius of Gregor Mendel, the Father of Genetics* (Boston: Houghton Mifflin, 2000), 96.

30 "Comments on Corpulency, Lineaments of Leanness, Mems on Diet and Dietetics": See William Wadd, *Mems on Diet and Dietetics* (London: John Ebers & Co., 1829).

31 "*was among the first to calculate the caloric values of food*": See listing for Baron von Justus, in *Random House Webster's Dictionary of Scientists* (New York: Random House, 1997), 298.

32 "'*saccharine and farinaceous diet*' *to fatten up farm animals*": See preface, William Banting, *A Letter on Corpulence, Addressed to the Public* (London: Harrison and Sons, 1869), 59, Pall Mall. Banting published four editions of his *Letter,* and in this, the final edition, he included a fourteen-page preface that detailed the influence of Bernard's talk on Harvey and Harvey's subsequent conception of the low-carbohydrate, high-protein regimen. Banting also points out in this preface that the diet was not entirely new, but had been previously prescribed not for the treatment of corpulence, but as a booster of athletic prowess.

32 "*I do not know of, nor can I imagine, any more distressing than that of Obesity*": See William Banting, *A Letter on Corpulence Addressed to the Public* (London: Harrison and Sons, 1863).

"*Clinical judgment relied on the elucidation of those senses, not necessarily on a systematic understanding of illness and its cure*": See Porter (1997), op. cit., 312.

33 "*thermometer of public health . . . involving the first systematic analyses of disease patterns*": See ibid., 406.

"*2,000 Frenchmen had lied about their height in an unsuccessful attempt to dodge the draft*": See essay by G. M. Jolly and P. Dagnelie in R. C. Olby, ed., *Early Nineteeth Century European Scientists* (London: Pergamon Press, 1967).

"*impossibility of defining absolutely what degree of obesity is to be considered morbid*": See John Forbes, ed. *Cyclopedia of Practical Medicine,* Volume III (Philadelphia: Lea and Blanchard, 1845).

35 "*Atwater managed to reset the nutritional standards for the nation*": Schwartz (1986), op. cit., 86–88.

"*the main facts will gradually work their way to the masses, who most need its benefit*": See W. O. Atwater, "Pecuniary Economy of Food," *Century Magazine* (January 1888): 437, 442.

36 "*Atwater of wishing to feed the U.S. laborer 'at a cost as low as Chinamen are subjected to'*": See Naomi Aronson, "Nutrition as a Social Problem," *Social Problems* 29 (5) (1982): 474–487.

"*present waste of food material could be spent for more adequate shelter, the bad tenements in the slums would be renovated*": See Aronson (1982), ibid.

"*serves in the body to either form tissue or yield energy, or both*": See W. O. Atwater and F. G. Benedict, "A Respiration Calorimeter with Appliances for the Direct Determination of Oxygen," Carnegie Institute of Washington, Washington, D.C. (1905).

"*the marks of which are to be detected on all sides and in no uncertain fashion*": See Russell H. Chittenden, *The Nutrition of Man* (Cambridge, UK: Cambridge University Press, 1907).

37 "*indication that excess food is being avoided*": Ibid.

"*quantities of food as customs and habits call for*": Ibid.

38 "*French and American dieting traditions*": See Peter N. Stearns, *Fat History: Bodies and Beauty in the Modern West* (New York and London: New York University Press, 1997).

"*Railing against men who were 'unduly stout'*": See Schwartz (1986), op. cit., pp. 80–82; Emma E. Walker, "Pretty Girl Papers," *Ladies' Home Journal,* January 1905, 33.

38 "*the century of svelte*": See William Bennett and Joel Gurin, *The Dieter's Dilemma: Eating Less and Weighing More* (New York: Basic Books, 1982). A wonderfully written, comprehensive, and rather radical book in its time, this book popularized the set-point theory of body weight stabilization.

39 "*waist cinching styles of the Romantic era*": See Nancy Etcoff, *Survival of the Prettiest: The Science of Beauty* (New York: Anchor Books, 2000). Etcoff makes clear that Poiret and other early-twentieth-century clothing designers liberated women from corsets, which made their figure flaws perhaps more evident, but cautions that at the time a flawless body was not the expectation.

40 "*domestic inefficiency in the midst of a floodtide of goods*": See Schwartz (1986), op. cit., 136.

"*of the remedy could be thrust into public view*": See *New York Times,* January 7, 1906.

"*that weighed not only bodies, but worthiness—the lower the number, the loftier the soul*": Christine Terhune Herrick, daughter of the widely read domestic-advice writer Marion Harland, was the Martha Stewart of her day, the author of dozens of books and articles focused on such salient concerns as "The Wastes of the Household: Watching and Saving the 'Left-Overs,'" which appeared in the inaugural issue of *Good Housekeeping* in 1885.

42 "*release through what one German psychoanalyst dubbed 'the alimentary orgasm'*": See S. Rado, "Uber die Psychische Wirkung der Rauschgifte," *Internationale Zeitschrift fur Psychoanalyse* 12 (1926), cited in Harold Kaplan and Helen Singer Kaplan, "The Psychosomatic Concept of Obesity," *Journal of Nervous and Mental Disease* 125 (1957): 181–201.

"*Psychiatrist Harold Kaplan and his wife clinical psychologist Helen Singer Kaplan*": The Kaplans later divorced and Helen Singer Kaplan went on to get her medical degree and to become a well-known sex therapist.

"*many symbolic meanings have been assigned to food*": See, for example, Kaplan (1957), op. cit.

"*sublimation or escape*": See, for example, "The Fat Personality," *Newsweek* (November 17, 1952): 110. The article points out that "Rich sauces . . . offer a delightful panacea for boredom, unhappy homes . . . sexual maladjustment."

"*physical expression of which is the accumulation of fat*": TAC Rennie, "Obesity as a Manifestation of Personality Disturbance," *Dis Nerv System* 1 (8) (1940).

44 "*objective test of validation*": See Hilde Bruch, "Transformation of Oral Impulses in Eating Disorders," *Psychiatric Quarterly* 35 (1961): 458–81.

45 "*had lower than average suicide rates*": See Hilde Bruch, *Eating Disorders: Obesity, Anorexia Nervosa, and the Person Within* (New York: Basic Books, 1973), 127–129.

46 "*with a weakness for gossip*": Thanks to my colleague, writer and author Mark Kramer, for sharing his remembrances of Stanley Schachter, whom he met while a graduate student at Columbia University.

"*to eat in response to environmental cues*": See, for example, Stanley Schachter and Larry Gross, "Manipulated Time and Eating Behavior," *Journal of Personality and Social Psychology* 10 (1968): 98–106; Stanley Schachter, Ronald Goldman, and Andrew Gordon, "Effects of Fear, Food Deprivation and Obesity on Eating," *Journal of Personality and Social Psychology* 10 (1968): 91–97.

47 "*Psychology of Successful Weight Control*": Mary Catherine Tyson and Robert Tyson (Chicago: Nelson Hall Co., 1974).

"*in 1997, Psychology Today*": See "Psychology Today's 1997 Body Image Survey Findings," *Psychology Today* (January/February 1997).

Chapter Three
Natural Born Freaks

49 John Locke, "Some Thoughts Concerning Education," edited with Introduction, Notes, and Critical Apparatus by John W. Yolton (Oxford University Press, 2000). This essay can also can be found in the "Internet Modern History Sourcebook," www.fordham.edu/halsall/mod/modsbook.html.

"*fourteen hundred pounds of flesh to his grave*": See, for example, *Guiness Book of World Records 1988* (New York: Bantam Books, Sterling Publishing Co., 1988), 10–12. While Minnoch was estimated to have been fourteen hundred pounds when he died, his highest recorded weight was actually one thousand pounds, after which he stopped weighing himself.

"*pound for pound, humans are by far the fattest*": See Caroline M. Pond, *The Fats of Life* (Cambridge, UK: Cambridge University Press, 1998). This is a wonderfully written evocation of the physiology, storage, and use of fats.

50 "*she would probably be too bulky to walk*": I gleaned a basic understanding of the working of fat cells from interviews with two world experts in the field: Joseph Vasselli, a biopsychologist whose primary research interest is mechanisms of appetite and body weight regulation at the Obesity Research Center, St. Luke's-Roosevelt Hospital Center, New York; and Susan Friedman, professor in the department of nutritional sciences, Cook College, Rutgers, whose research focus is on regulation of human adipose tissue metabolism and function.

52 "*determined the inheritance of traits*": A good deal of what I know of Mendel derives from my reading of Robin Marantz Henig's lucid and lyrical biogra-

phy, *The Monk in the Garden* (Boston: Houghton Mifflin, 2001). For a captivating semifictionalized account of the man and his work see Simon Mawer, *Mendell's Dwarf* (New York: Harmony Books, 1998).

52 "*chemical recipe for a single protein*": See A. E. Garrod, "The Incidence of Alkaptonuria: A Study in Clinical Individuality," *Lancet* 2 (1902): 1616–1620.

53 "*chemical individuality as the paradigm of Mendelian variation*": See A. E. Garrod, "The Croonian Lectures on Inborn Errors of Metabolism, Lecture II: Alkaptonuria," *Lancet* 2 (1908): 73–79.

"*four nucleic acid bases: adenine, cytosine, guanine and thymine, A, C, G, and T*": Bacteriologist Oswald Theodore Avery laid the groundwork for modern genetics and molecular biology by proving conclusively in 1944 that DNA from the nucleus of the cell is the genetic material. Avery spent most of his research life at Rockefeller Institute where he studied pneumococcus, a particularly deadly bacterium that causes lobar pneumonia. He identified DNA as the substance of the genes in a 1944 paper coauthored by Colin MacLeod and Maclyn McCarty: "Studies on the Chemical Nature of Substance Inducing Transformation of Pneumococcal Typoes. Induction of Transformation by a Desoxyribonucleic Acid Fraction Isolated from Pneumococcus Type III," *Journal of Experimental Medicine* 79: 137–158. His finding—that DNA is the hereditary material—is the seminal biological insight of the twentieth century.

"*twenty-three pairs of chromosomes are volumes*": The literature on genes and genomics is enormous, and I have drawn on a number of texts, as well as interviews, to flesh out the history. Among the more informative books on the topic are Steven Jones, *The Language of the Genes* (New York: Harper Collins, 1993), and David Micklos and Gregory Freyer, *DNA Science* (Cold Spring Harbor Laboratory Press, 1990). A recent popular account by Matt Ridley, *Genome: An Autobiography of a Species in 23 Chapters* (New York: HarperCollins, 1999), is useful for its metaphors and explanations.

54 "*breast cancer gene never contract the disease*": For a very interesting discussion of the limitations of the gene as a metaphor, consider Evelyn Fox Keller, *A Century of the Gene* (Harvard University Press, 2000). Harvard biologist Richard Lewontin has also written voluminously and well on this theme. For a good overview of his insights see his collection of essays, *Ain't Necessarily So: The Dream of the Human Genome and Other Illusions* (New York Review of Books, 2000).

56 "*analysis of their inheritance and genetic variation*": See Lee M. Silver, *Mouse Genetics: Concepts and Applications* (New York: Oxford University Press, 1995). This book is available on-line at www.informatics.jax.org/silver/index.shtml.

56 "*in the degree to which cancer appears and the type of cancer which arises*": See Jean Holstein, *The First Fifty Years at the Jackson Laboratory*, The Jackson Laboratory, 1979.

57 "*incinerating all but a scattering of the mice*": Fire, as it turned out, would be an ongoing problem at Jackson Laboratories into the early 1990s, and the resulting importation of mice from other labs would continue to introduce new mutants to the facility.

"*double the weight of the ordinary variety*": Lethal yellow is an obesity syndrome in mice associated with diabetes and insulin resistance and a yellow coat color due to the widespread expression in the mouse of the agouti protein. See S. Bultman, E. Michaud, and R. Woychik, "Molecular Characterization of the Mouse Agouti Locus," *Cell* 71 (1992): 1195–1204.

60 "*the mice died*": Douglas L. Coleman and Katharine P. Hummel, "Effects of Parabiosis on Normal with Genetically Diabetic Mice," *American Journal of Physiology* 217 (November 1969): 1298–1304.

61 "*in such a way that feeding is reduced*": See G. R. Hervey, "The Effects of Lesions in the Hypothalamus in Parabiotic Rats," *Journal of Physiology* 145 (1959): 336–352.

"*christened a 'lipostat'*": See Gordon C. Kennedy, "The Role of Depot Fat in the Hypothalmic Control of Food Intake in the Rat," *Proceedings of the Royal Society* 140 (1953): 578–592.

62 "*steady state or 'set-point'*": For a good summary of this widely published, remarked upon, and studied theory, see R. E. Keesey, "A Set-point Theory of Obesity," in K. D. Brownell and J. P. Forety, eds., *Handbook of Eating Disorders: Physiology, Psychology and Treatment of Obesity, Anorexia and Bulimia* (New York: Basic Books, 1986), 63–87.

"*but lack the equipment—the receptor—to sense it*": See D. L. Coleman, "Effects of Parabiosis of Obese with Diabetes and Normal Mice," *Diabetologia* 9 (1973): 294–298.

63 "*creates a new orthodoxy*": Thomas S. Kuhn, *The Structure of Scientific Revolutions* (University of Chicago Press, 1962).

64 "*live in a semistarved state*": See A. J. Stunkard, Presidential Address, 1974, "From Explanation to Action in Psychosomatic Medicine: The Case of Obesity," *Psychosom Med* 37 (3) (May–June 1975): 195–236.

"*The results, they politely concluded, were 'disappointing'*": See Albert Stunkard and Sydnor Penick, "Behavior Modification in the Treatment of Obesity," *Archives of General Psychiatry* 36 (1979): 801–806.

"*'Two Tramps in Mud Time'*": See *Complete Poems of Robert Frost* (New York: Henry Holt, 1949).

CHAPTER 4
ON THE CUTTING EDGE

67 "*he wasn't allowed to do something, he did it*": Telephone interview with Scott Friedman, April 8, 2001.

"*not a standout academically, a bit of a joker*": Telephone interview with George Schilling, Jeffery Friedman's high school biology teacher, June 14, 2001.

68 "*Full professors here are treated like feudal lords*": See Barbara J. Culliton, "Rockefeller Braces for Baltimore," *Science* 247 (January 12, 1990): 148–151.

69 "*retained a mystic faith that God had given him money for mankind's benefit*": Ron Chernow, *Titan: The Life of John D. Rockefeller, Sr.* (New York: Random House, 1998), 468.

"*horrified to learn how few bacterial diseases were within the power of physicians to treat*": See Roy Porter, *The Greatest Benefit to Mankind* (New York: W. W. Norton, 1997), 531.

"*then at least satisfaction on earth*": Frederick Gates later sent Osler a letter of appreciation, in which he wrote: "Some years ago, in carrying out a determination to become more intelligent as a layman on the subject of the current and common diseases, I purchased a copy of your *Principles and Practice of Medicine,* on the advice of a bright young medical friend. Happening to receive it just as I was to start on a vacation, I took the book with me and read it from beginning to end, with absorbing interest, and with a medical dictionary at my side.

"In reading it I was impressed especially with the vast numbers of diseases that are certainly or probably originated by bacteria . . . and the vast possibilities for good lying in this field of research opened up before my imagination and fired my enthusiasm. I, therefore, laid the matter before Mr. Rockefeller, and sought to impart to him my own interest, kindled by the reading of your book, in bacteriological research. His enthusiasm was easily kindled . . . and the result was the Rockefeller Institute, of which you remember Dr. Welch is the Chairman, with an initial and tentative working fund of $200,000 with which to experiment . . . Mr. Rockefeller contributed a million dollars to the Harvard School [because of] the very superior work done at that institution . . . Both of these gifts grew directly out of your book. It has occurred to me that possibly you might be gratified to know of an incidental and perhaps to you quite unexpected good which your valuable work has wrought."

70 "*was likely at any moment to be precipitated*": See Gerald M. Edleman, "Basic Research the Need for Knowledge," quoting from a 1910 report by Abraham Flexner. Selection from *Beyond Tomorrow: Trends and Perspectives in Medical Science* (New York: The Rockefeller University, 1977).

70 "*explore and dream*": See Porter (1997), op. cit.

72 "*Coleman's mysterious satiety factor, it seemed, had been found*": See Eugene Straus and Rosalyn S. Yalow, "Cholecystokinin in the Brains of Obese and Nonobese Mice," *Science* 203 (January 5, 1979): 68–69.

"*It could not be* the *satiety factor absent in* ob *mice*": This account was built from interviews with Bruce Schneider over a period of months. For a detailed and highly authoritative account of the early work on CCK and satiety, see Robert Pool, *Fat: Fighting the Obesity Epidemic*" (New York: Oxford University Press, 2000).

73 "*I published my findings in the* Journal of Clinical Investigation": See Bruce S. Schneider, Joseph W. Monahan, and Jules Hirsh, "Brain Cholecystokinin and Nutritional Status in Rats and Mice," *Journal of Clinical Investigation* 64 (November 1979): 1348–1356.

Chapter 5
Hunger

77 "*where the firelight doesn't reach*": Hayden Carruth, "Notes on Empysema," *Harper's Magazine* (October 2001): 28–29.

80 "*It was later discovered that both of these men had a family history of obesity*": See E. A. H. Sims and E. S. Horton, "Adaptation to Obesity and Starvation," *American Journal of Clinical Nutrition* 21 (1968): 1455–1470; and E. A. H. Sims, R. F. Goldman, C. M. Gluck, E. S. Horton, P. C. Kelleher, and D. W. Rowe, "Experimental Obesity in Man," *Transactions of the Association of American Physicians* 81 (1968): 153–170.

81 "*I have no more sexual feeling than a sick oyster*": See Ancel Keys, *The Biology of Human Starvation,* Volume II (The University of Minnesota Press, 1950), 839.

"*stomach masturbation*": See Hilde O Bluhm. "How Did They Survive?" *American Journal of Psychotherapy* 2 (1) (1948): 20.

82 "*I envy the fat pigeons picking at them*": Keys (1950), op. cit., 852.

"*theory of weight stabilization that had been bandied about for nearly a century*": See Robert Pool, *Fat: Fighting the Obesity Epidemic* (New York: Oxford University Press, 2001), 73.

"*Gordon Kennedy's 'lipostat'*": See Gordon C. Kennedy, "The role of depot fat in the hypothalmic control of food intake in the rat," *Proceedings of the Royal Society* 140 (1953): 578–592.

"*maintain a steady state or 'set-point'*": R. E. Keesey, "A set-point theory of obesity," in K. D. Brownell and J. P. Forety, eds., *Handbook of Eating Disor-*

ders: Physiology, Psychology and Treatment of Obesity, Anorexia and Bulimia (New York: Basic Books, 1986), 63–87.

83 "*on the connection between body fat and fertility*": Rose E. Frisch and Janet W. McArthur, "Menstrual Cycles: Fatness as a Determinant of Minimum Weight for Height Necessary for their Maintenance or Onset," *Science* 13 (185) (September 1974): 949–951.

84 "*size or the number of fat cells, or both*": See, for example, Bruce M. Spiegelman and Jeffrey S. Flier, "Adipogenesis and Obesity: Rounding Out the Big Picture," *Cell* 87 (November 1, 1996): 377–389.

"*fat cells can and do sprout and die throughout life*": In 1972, Hirsh and his colleague Jerome L. Knittle famously proposed that the number of fat cells in humans was fixed in early childhood, and almost certainly by one's fifth birthday. This theory turned out to be incorrect, but it continues to influence thinking and discussion on the subject. I spoke with Bruce Spiegelman and Jeffrey Flier at length on the subject of fat cell growth and proliferation in humans, as well as with other scientists. What became clear is that due in part to the difficulty of studying cell growth in living human subjects, theories in this area remain somewhat controversial, though there is consensus that new fat cell growth is possible well into adulthood, and perhaps throughout life.

"*substantial increase in the number of fat cells*": Interview with researcher Susan Fried, professor in the Department of Nutritional Science at Rutgers University.

85 "*obstinate, curious or careless, quick or slow*": See John Locke, "Some Thoughts Concerning Education," ed. John W. Yolton (New York: Oxford University Press, 2000). This essay can also can be found in the "Internet Modern History Sourcebook" at www.fordham.edu/halsall/mod/modsbook.html.

86 "*hard-wired into the organism, and therefore out of mindful control*": For a thoughtful discussion of the "atom of behavior" theory in the context of early thinking on the biological basis of behavior, see Jonathan Weiner, *Time, Love, Memory* (New York: Knopf, 1999). Weiner's book is among the best—and certainly the best written—accounts of this fascinating and controversial field.

87 "*within one's own soul*": C. G. Jung, *Man and His Symbols* (London: Aldus, 1964), 10.

88 "*the significance of genetics in human obesity*": See, for example, A. J. Stunkard, "Genetic Contributions to Human Obesity," *Res Publ Assoc Res Nerv Ment Dis* 69 (1991): 205–218; A. J. Stunkard, J. R. Harris, N. L. Pedersen, and G. E. McClearn, "The body-mass index of twins who have been reared apart," *New England Journal of Medicine* 322 (21) (May 24): 1522–1524. R. R. Fabsitz, P. Sholinsky, and D. Carmelli, "Genetic Influences on Adult Weight Gain and

Maximum Body Mass Index in Male Twins," *American Journal of Epidemiology* 140 (8) (1994): 711–720. A. J. Stunkard et al., "A Twin Study of Human Obesity," *JAMA* 256 (1986): 51–54; H. H. M. Maes et al., "Genetic and Environmental Factors in Relative Body Weight and Human Adiposity," *Behavior Genetics* 27 (4) (1997). See also Claude Bouchard et al., "The Response to Long-Term Overfeeding in Identical Twins," *New England Journal of Medicine* 322 (21) (May 24, 1990): 1477–1482.

88 "*raised far apart in different adoptive families*": Claude Bouchard, now executive director of the Pennington Biomedical Research Center in Baton Rouge, Louisiana, has done seminal work on the genetics of *ob*esity, and in an interview outlined the history of twin studies and their role in elucidating the genetics of *ob*esity.

91 "*father of American genetics*": See Matt Ridley, *Genome* (New York: HarperCollins, 2000), 45.

"*nineteen-year-old Alfred Sturtevant*": For a detailed account of Morgan's work, as well as of the days of early-twentieth-century fruit fly genetics and the genetics of behavior, see Jonathan Weiner (1999), op. cit. For an affectionate and highly entertaining look at the early players in the game of fly genetics, see Robert E. Kohler, *Lords of the Fly: Drosophila Genetics and the Experimental Life* (University of Chicago Press, 1994).

"*the gene for a type of muscular dystrophy*": See M. Koenig, E. P. Hoffman, C. J. Bertelson, A. P. Monaco, C. A. Feener, and L. M. Kunkel, "Complete Cloning of the Duchenne Muscular Dystrophy (DMD): cDNA and Preliminary Genomic Organization of the DMD Gene in Normal and Affected Individuals," *Cell* 50 (1987): 509–517.

92 "*lead to the discovery of the Huntington's gene*": See J. F. Gusella et al., "A Polymorphic DNA Marker Genetically Linked to Huntington's Disease," *Nature* 306 (1983): 234–238.

"*prepared increasingly more refined genetic maps*": See N. Bahary, R. L. Leibel, L. Joseph, and J. M. Friedman, "Molecular Mapping of the Mouse db Mutation," *Proc Natl Acad Sci* (1990): 8642–8646; and J. M. Friedman, R. L. Leibel, D. S. Siegel, J. Walsh, and N. Bahary, "Molecular Mapping of the Mouse *ob* Mutation," *Genomics* 11 (1991): 1054–1062. Never in any of his scores of written or verbal communications with me did Jeffrey Friedman mention Nathan Bahary, or credit him for this masterful—and well-documented—achievement.

94 "*a technical tour de force, just beautiful*": I first asked Dr. Friedman about his collaboration with Dr. Leibel at a meeting we had at Rockefeller University on April 9, 1999. He told me that their parting had been "amicable." When I later

asked for clarification of this, he became visibly agitated and declined to elaborate. When in still another meeting I asked Friedman for the names of his collaborators on the initial cloning of the gene, he insisted that he alone was responsible for the intellectual content of the work, and others in attendance on the project were mere "technicians."

94 "*essentially staying out of the lab where* ob *was being cloned*": Dr. Friedman never made mention of Dr. Zhang's or of Dr. Seigel's contribution in any written or verbal communication with me, though their efforts were certainly central to the discovery.

96 "*Maffei was at a friend's wedding that weekend*": E-mail communication with Dr. Margherita Maffei, Dipartimento di Endocrinologia e Metabolismo Ospedale di Cisanello, June 5, 2001.

97 "*with fewer than forty-four seconds on the clock*": For accounts of the Knicks/Bulls game see A. Harvin, *New York Times,* May 9, 1994, C8, and Clifton Brown, *New York Times,* May 8, Section 8, 12. For background on Pete's Tavern see Whitney Walker, *New York Daily News,* March 6, 1996, A1.

98 "*learning everything they could about the* ob *gene and its mutations*": For a careful account of the cloning of the *ob* gene as told from Jeffrey Friedman's perspective, see Robert Pool (2001), op. cit. Pool's scientific account of this period is quite detailed. He relates the tale precisely as Friedman tells it, giving only passing credit for the discovery to other researchers with links to the laboratory.

99 "*Friedman insisted on the separate publication in 1993 of the microdissection results*": See Nathan Bahary, Donald A. Siegel, Justine Walsh, Yiying Zhang, Lori Leopold, Rudolph Leibel, Ricardo Proenca, and Jeffrey M. Friedman, "Microdissection of Proximal Mouse Chromosome 6: Identification of RFLPs Tightly Linked to the *ob* Mutation," *Mammalian Genome* 4 (1993): 511–515.

100 "*large amount of active material necessary for animal experiments*": Thanks to Stephen Burley for his patient and thorough explanation of this challenging scientific effort in his e-mail communication to me on June 21, 2001.

101 "*lost every gram of their body fat*": J. L. Halaas, K. S. Gajiwala, M. Maffei, S. L. Cohen, B. T. Chait, D. Rabinowtiz, R. L. Lallone, S. K. Burley, and J. M. Friedman, "Weight Reducing Effects of the Plasma Protein Encoded by the *ob*ese Gene," *Science* 269 (1995): 543–546.

"*tested leptin and reported similar results*": See L. A. Campfield, F. J. Smith, Y. Guisez, R. Devos, and P. Burn, "Recombinant Mouse *OB* protein: Evidence for a Peripheral Signal Linking Adiposity and Central Neural Networks," *Science* 269 (1995): 546–549. Thanks also to Arthur Campfield for giving me a firsthand account of this in our meeting in Long Beach, California, in October 2000.

101 "*Soon thereafter, several other laboratories tested leptin and reported similar results*": See, for example, M. A. Pelleymounter, M. J. Cullen, M. B. Baker, R. Hecht, D. Winters, T. Boone, and F. Collins, "Effects of the *obese* Gene Product on Body Weight Regulation in *ob/ob* Mice," *Science* 269 (1995): 540–543.

"*'into my eyeball,' if necessary*": William Haseltine, chairman of the board of directors and chief executive officer of Human Genome Sciences Inc., recalled being told this by Steingarten in a conversation we had during a private breakfast meeting at the Ritz Carlton in Boston, May 1999.

102 "*reportedly the largest ever for a university-held patent*": Robert Pool (2001), op. cit. 130.

CHAPTER 6
THE CLINICAL EXCEPTION

105 "*The obese is . . . in a total delirium*": Jean Baudrillard, "Figures of the Transpolitical," in Jim Fleming, ed., *Fatal Strategies,* translated by Philip Beitchman and W.G.J. Niesluchowski (New York: Semiotext; London: Pluto, 1990).

"*I'm fat, but I'm thin inside*": George Orwell, "Coming Up for Air" (1939), Peter Davison, ed., assisted by Ian Angus and Sheila Davison, *The Complete Works of George Orwell* (London: Secker & Warburg, 1986), vol. 7.

"*But in the industrialized West, obesity and poverty are to some degree linked, especially in women*": See Albert Stunkart and T. L. Sorenson: "Obesity and Socioeconomic Status: A Complex Relation," *New England Journal of Medicine* 329 (14) (September 30, 1993): 1036–1037. See Nancy Fultz et al., "Economic Penalty of Extra Pounds to Middle Aged Women," paper presented at the annual meeting of the Gerontological Society of America, November 10, 2000. The relationship between income and obesity has begun to weaken in the affluent industrialized nations, as the pandemic sweeps an increasing number of people in. For example, for Caucasian men in the United States obesity rates are not linked to income level.

107 "*a luxury that the poor and working classes either cannot access or cannot afford*": Tomas J. Philipson and Richard A. Posner, "The Long-Run Growth in Obesity as a Function of Technological Change," National Bureau of Economic Research, Working Paper #7423, November 1999, available at www.nber.org/papers/w7423. Philipson, an economist, and Posner, a federal appeals court judge and prolific author, make the fascinating argument that while exercise in the form of manual labor at one time provided a source of income, it now entails a cost in the form of lost leisure time. They argue as well that the sharp increase in paid labor among women in the last three decades may contribute to the rise of obesity and overweight in women, because housework is more strenuous than sitting at a desk.

110 "*what he called a 'semi-starved neurosis'*": Ancel Keys, *The Biology of Human Starvation*, volume II (Minneapolis: University of Minnesota Press, 1950), 908.

"*shifting just out of reach*": Ejnar Mikkelsen, *Lost in the Arctic: Being a Story of the Alabama Expedition* (London: William Heinemann, 1913), 270.

111 "*complete degeneration of character as chronic starvation*": G. B. Leyton, "A Survey of the Effect of Slow Starvation on Man," thesis, Cambridge University (1946).

112 "*And tests showed that many of them did*": Robert V. Considine et al., "Serum Immunoreactive-Leptin Concentrations in Normal Weight and Obese Humans," *New England Journal of Medicine*, 334 (February 1, 1996): 292–295.

114 "*Equally important, the Amgen volunteers were adults*": S. B. Heymsfield, A. S. Greenberg, K. Fujioka, R. M. Dixon, R. Kushner, T. Hunt, J. A. Lubina, J. Patane, B. Self, P. Hunt, and M. McCamish, "Recombinant Leptin for Weight Loss in Obese and Lean Adults: A Randomized, Controlled, Dose-escalation Trial," *JAMA* 282 (16) (October 27, 1999): 1568–1575.

115 "*announced their discovery in 1997*": See Carl T. Montague, I. Sadaf Farooqi, et al., "Congenital Leptin Deficiency is Associated with Severe Early-Onset Obesity in Humans," *Nature* 387 (June 26, 1997).

"*weighed 330 pounds and had a fifty-five-inch waist*": See A. Strobel, T. Issad, L. Camoin, M. Ozata, and A. D. Strosberg, "A Leptin Missense Mutation Associated with Hypogonadism and Morbid Obesity," *Nature Genetics* (March 2, 1998).

116 "*a reduction of fertility—is protective*": For an excellent review of leptin and its impact on body weight, fertility, and immune response, see J. M. Friedman and J. L. Halaas, "Leptin and Regulation of Body Weight in Mammals," *Nature* 395 (1998): 763–770.

"*The publication of this finding in the* Journal of the American Medical Association": Steven B. Heymsfield et al. (October 27, 1999), op. cit.

117 "*that some subgroup of people may benefit from leptin injections*": I. Sadaf Farooqi et al., "Partial leptin deficiency and human adiposity," *Nature* 414 (6859) (November 1, 2001): 34–35.

"*probably not in many and certainly not in the majority*": See, for example, Marilyn Chase, "Early Human Trials Show Leptin May be Effective Drug," *Wall Street Journal*, June 15, 1998.

"*among people with identical percentages of body fat*": See Marina Chicurel, "Whatever Happened to Leptin," *Nature* 404 (April 6, 2000): 538–540.

"*have levels that are higher than those of the obese*": See M. Maffei et al., *Nature* Medicine 1 (1995): 1155–1161.

119 "*that had earlier lost the* ob *patent to Amgen*": See L. A. Tartaglia, M. Dembski, X. Weng, N. Deng, J. Culpepper, R. Devos, G. J. Richards, L. A. Campfield, F. T. Clark, J. Deeds, et al., "Identification and expression cloning of a leptin receptor, OB-R," *Cell* 83 (7) (December 29, 1995): 1263–1271.

"*subsequent patenting was sweet revenge*": While the Millennium team was the first to identify a leptin receptor in the 1995 paper in *Cell*, Friedman's team at Rockefeller published the more definitive paper, describing the full-length leptin receptor and its mutation in the *db* mouse. See G. H. Lee, R. Proenca, J. M. Montez, K. M. Carroll, J. G. Darvishzadeh, J. I. Lee, and J. M. Friedman, "Abnormal Splicing of the Leptin Receptor in Diabetic Mice," *Nature* 379 (6566) (February 15, 1996): 632–635.

"*As Rudy Leibel and others proved*": See S. C. Chua, W. K. Chung, X. S. Wu-Peng, Y. Zhang, S. M. Lui, L. Tartaglia, and R. L. Leibel, "Phenotypes of Mouse *Diabetes* and Rat *Fatty* Due to Mutations in the OB (Leptin) Receptor," *Science* 271 (1996): 994–996.

"*So far only three people are known to lack the receptor*": See K. Clement, C. Vaisse, N. Lahlou, S. Cabrol, V. Pelloux, D. Cassuto, M. Gourmelen, C. Dina, J. Chambaz, J. Lacorte, A. Basdevant, P. Bourgnere, Y. Lebouc, P. Froguel, and B. Guy-Grand, "A Mutation in the Human Leptin Receptor Gene Causes Obesity and Pituitary Dysfunction," *Nature* (1998): 392–398.

"*scientists are working hard to find whether this is indeed the case*": See, for example, J. H. Lee et al., "Leptin Resistance Is Associated with Extreme Obesity and Aggregates in Families," *International Journal of Obesity* 25 (2001): 1471–1473.

121 "*it was a symptom of a rare genetic syndrome*": See Robert S. Jackson et al., "Obesity and Impaired Prohormone Processing Associated with Mutations on the Human Prohormone Convertase 1 Gene," *Nature Genetics* 16 (July 1997): 303.

"*in a* Nature Genetics *editorial*": Rudolph Leibel, "And Finally, Genes for Human Obesity," *Nat Genet* 3 (July 16, 1997): 218–220.

122 "*a protein made from POMC that reduces appetite*": G. S. Yeo et al., "A Frameshift Mutation in MC4R Associated with Dominantly Inherited Human Obesity," *Nat Genet* 20 (1998): 111–112

"*A related defect had been identified years earlier in the agouti mouse*": Roger Cone, a senior scientist at the Vollum Institute at the Oregon Health and Sciences University in Portland, did most of the fundamental work on the melanocortin receptors. In an E-mail correspondence to me he explained that after cloning the various melanocortin receptors, his team sought to determine their function, and to determine how the lethal yellow allele of agouti caused both yellow pigmentation and obesity, since the yellow coat color implicated a melanocortin receptor. His first breakthrough was the demonstration that the agouti peptide was a competitive antagonist not only of the MC1-R, but also of the MC4-R [Lu et al., *Nature* 371 (1994): 799–802]. In this paper, he proposed that expression of agouti peptide in the lethal yellow agouti mouse blocked the central MC4-R to cause the agouti obesity syndrome. To test this, his team chose to both derive a MC4-R antagonist and to knock out the gene. They then found the first MC4-R antagonist [Hruby et al., *J Med Chem* 38 (1995): 3454–3461], and went on to use this to prove their hypothesis that the central melanocortin system is involved in the regulation of energy homeostasis [Fan et al., *Nature* 385 (1997): 165–168]. Cone wrote: "Went I first visited Millenium, they were trying to clone an agouti receptor from adipocytes, as they believed agouti acted on adipocytes to cause obesity. I convinced them otherwise with my pharmacological data on agouti from Lu et al., and they then informed me that, despite the fact I already had a MC4-R knockout program, they would start their own, but that they also wanted to collaborate. They ended up beating us fair and square, getting the knockout first. However, our agreement was to collaborate on the paper, irrespective of who won the race." That paper was D. Huszar, C. A. Lynch, V. Fairchild-Huntress, J. H. Dunmore, Q. Fang, L. R. Berkemeier, W. Gu, R. A. Kesterson, B. A. Boston, R. D. Cone, F. J. Smith, L. A. Campfield, P. Burn, and F. Lee, "Targeted Disruption of the Melanocortin-4 Receptor Results in Obesity in Mice," *Cell* 88 (1) (January 10, 1997): 131–141.

123 "*making it the most common genetic defect yet to be associated with obesity*": See, for example, I. S. Farooqi, G. S. Yeo, J. M. Keogh, S. Aminian, S. A. Jebb, G. Butler, T. Cheetham, and S. O'Rahilly, "Dominant and Recessive Inheritance of Morbid Obesity Associated with Melanocortin 4 Receptor Deficiency," *J Clin Invest* 106 (2) (July 2000): 271–279; and Serge Hercberg, Bernard Guy-Grand, and Philippe Froguel, "Melanocortin-4 Receptor Mutations are a Frequent and Heterogeneous Cause of Morbid Obesity," *J Clin Invest* 106 (2) (July 2000): 253–262. The work of Philippe Froguel and his laboratory has been critically important in the sorting out of these mutations. Among many other accomplishments, Froguel's lab has been the only one to locate the leptin receptor mutation in humans.

"*that mice without an active form of the MCH gene*": Genes are "knocked out," or bred out of laboratory mice and sometimes other animals by scientists. Because the genetic patterns of mice and humans are similar, knockout mice provide valuable models of human diseases.

"*In the spring of 2001, Jeff Friedman announced in* Science": J. De Falco, M. Tomishima, H. Liu, C. Zho, X. Cai, J. D. Marth, L. Enquist, and J. M. Friedman, "Virus Assisted Mapping of Neural Inputs to a Feeding Center in the Hypothalmus," *Science* 291 (2001): 2608–2613.

Chapter 7
Collateral Damage

126 "*are more likely to enjoy the continued support of the makers of those products*": Predictably, company-funded research tends to favor company products. For example, a 1998 report in the *New England Journal of Medicine* revealed that 96 percent of medical journal authors whose research was favorable to certain cardiac drugs had financial ties to the drugs' manufacturers. By contrast, only 37 percent of authors of studies critical of those drugs had such financial ties. And according to a 1996 study published in the *Annals of Internal Medicine,* an amazing 98 percent of company-sponsored drug studies published between 1980 and 1989 in peer-reviewed journals or in symposia proceedings favored the funding company's drug. See H. T. Stelfox et al., "Conflict of Interest in the Debate Over Calcium-Channel Antagonists," *New England Journal of Medicine* 338 (2) (January 8, 1998): 101–106, and M. K. Cho and L. A. Bero, "The Quality of Drug Studies Published in Symposium Proceedings," *Annals of Internal Medicine* 124 (5) (March 1, 1996): 485–489.

"*the fat substitute made by Proctor & Gamble*": J. C. Lovejoy et al., "Beneficial Effect of a Low-Fat Diet on Health Risk Factors Is Mediated by Weight-Loss in Middle-Age Men," *Obesity Research* (8) Suppl 1 (October 2000).

"*what turned out to be the most lavish scientific session of all, on its weight loss preparation, Meridia*": In December 2000, Chicago-based Abbott Laboratories paid $6.9 billion for the Knoll Pharmaceutical unit of BASF AG, the European conglomerate that previously owned the company. Knoll was expected to post $2.1 billion in sales that year, thanks largely to sales of Meridia and its thyroid drug, Synthroid.

127 "*on the list enjoyed one or more industry affiliations*": See Susan Z. Yanovski, "Development of Eating Disorders in Overweight and Obese Adults," *Archives of Internal Medicine* 160 (17) (September 25, 2000): 2581–2589.

"*Fourteen percent of these relationships involved equity ownership*": See Elizabeth A. Boyd and Lisa A. Bero, "Assessing Faculty Financial Relationships With Industry: A Case Study," *JAMA* 284 (17) (November 1, 2000).

"*According to a 1996 study in the* Annals of Internal Medicine": Op. cit., M. K. Cho and L.A. Bero (1996).

"*and, of course, Knoll Pharmaceuticals Ltd.*": *Wall Street Journal,* February 9, 1998, B1.

128 "*thyroid medication, two of them generic forms that cost half as much*": See B. J. Dong, W. W. Hauck, J. G. Gambertoglio, L. Gee, J. R. White, J. L. Bubp, and F. S. Greenspan, "Bioequivalance of Generic and Brand-Name Levothyroxine Pro-ducts in the Treatment of Hypothyroidism," *JAMA* 277 (1997): 1205–1213.

128 "*Knoll tried mightily to suppress Dong's findings, and the eventual exposure of this landed the company in an embarrassing class action lawsuit*": For an eye-opening exposé of this and other issues surrounding the cozy relationship between academia and the drug industry, see David Shenk, "Money+Science= Ethics Problems on Campus," *The Nation* (March 22, 1999).

"*why Knoll would be foolish enough to commission such a 'no-win' study with predictable adverse marketing potential*": See Eugene Straus, "How Medical Research Becomes Advertising," Earth Times News Service.

"*U.S. sales of the drug soared to more than $108 million in 1998*": See Quintiles Transnational press release, "Viagra Leads in Retail Sales for '98 Drug Launches," June 11, 1999, which credits as its source, Scott-Levin, SourceTM Prescription Audit.

"*Knoll blamed the decline in part on a lack of 'consumer awareness' and launched a $50 million advertising campaign*": See "Meridia Gets $50 million launch effort," *Advertising Age*, October 21, 1998.

"*FDA allowed pharmaceutical companies to name both the drug and the disease in the same commercial without reeling off a long list of side effects*": See the FDA Modernization Act of 1997, available on-line at www.fda.gov/cder/guidance/105–115.htm.

"*sales of Meridia dropped once again, to a disappointing $94 million*": The weight loss industry is shrouded in secrecy, but Marketdata, a marketing research firm, has kept close tabs on it since 1989. Their reports on the industry were helpful, but perhaps more so were interviews with the company research director, John LaRosa, who was kind enough to share the obesity drug sales figures with me directly.

129 "*excluding asthmatics will make the task of finding study subjects more difficult*": One of these centers was Tom Wadden's clinic at the University of Pennsylvania, and the study was headed by Dr. Wadden's frequent coauthor, Robert Berkowitz.

130 "*as unhealthy, linking it to an increased risk of cardiovascular disease*": P. Hamm, R. B. Shekelle, and J. Stamler, "Large fluctuations in body weight during young adulthood and the twenty-five-year risk of coronary disease in men," *Amer J Epidemiol* 129 (1989): 312–318.

"*increased risk of cardiovascular disease, and a decline in lean body mass*": See, for example, T. Kajioka, S. Tsuzuku, H. Shimokata, and Y. Sato, "Effects of Intentional Weight Cycling on Non-Obese Young Women," *Metabolism* 51 (2) (February 2002): 149–154; and M. B. Olson, S. F. Kelsey, V. Bittner, S. E. Reis, N. Reichek, E. M. Handberg, and C. N. Merz, "Weight Cycling and High-Density Lipoprotein Cholesterol in Women: Evidence of an Adverse Effect:

A Report From the NHLBI-Sponsored Women's Ischemia Syndrome Evaluation Study Group," *J Am Coll Cardiol* 36 (5) (November 1, 2000): 1565–1571.

130 "*weight loss, if any, is likely to be temporary*": Weight cycling continues to be a concern. See, for example, A. Kroke, A. D. Liese, M. Schulz, M. M. Bergmann, K. Klipstein-Grobusch, K. Hoffmann, and H. Boeing, "Recent Weight Changes and Weight Cycling as Predictors of Subsequent Two Year Weight Change in a Middle-aged Cohort," *Int J Obes Relat Metab Disord* 26 (3) (March 2002): 403–409. The authors of this study conclude that a history of weight cycling is the "strongest predictor of subsequent large weight gain."

"*Bray ascribed this sorry state of affairs in part to a disappointing performance in the marketing end*": George Bray, keynote speech, NAASO Annual Meeting, Long Beach, California, October 2000.

131 "*Bray published articles written by him and by others extolling the virtues of Meridia*": See D. H. Ryan, P. Kaiser, and G. A. Bray, "Sibutramine: a novel new agent for obesity treatment," *Obes Res* 3 Suppl 4 (November 1995): 553S-559S.

"*in which he recommends that medication be 'seriously considered' for all overweight and obese patients over age ten*": See George A. Bray, *Contemporary Diagnosis and and Management of Obesity* (Newton, Penn.: Handbooks in Health Care Company, 1998), 157.

"*He does not believe that gaining and losing weight repeatedly is a particular danger, or, as others have argued, psychologically damaging*": See A. J. Stunkard, "Psychosocial Factors and Quality of Life in Obesity," in "The Health and Socio-Economic Costs of Obesity," Satellite Symposium to the 6th European Congress on Obesity.

"*any more than the occasional stroke after cessation of hypertensive therapy should block the use of anti-hypertensive drugs*": See Abbott Press Release, January 16, 2002.

"*Nor is there evidence that Meridia or any weight loss medication is safe over the long haul*": See P. J. Carek and L. M. Dickerson, "Current Concepts in the Pharmacological Management of Obesity," *Drugs* 57 (6) (June 1999): 883–904.

"*This was troubling, because weight loss is one of the most effective methods of* reducing *blood pressure*": See, for example, C. D. Mulrow, E. Chiquette, L. Angel, J. Cornell, C. Summerbell, B. Anagnostelis, et al., "Dieting to Reduce Body Weight for Controlling Hypertension in Adults," *Cochrane Database Syst Rev* 2000; (2): CD000484.

132 "*Citizen petitioned the federal government to take Meridia off the market, citing nineteen deaths linked to the drug*": See Philip J. Hilts, "Petition Asks for Removal of Diet Drug from Market," *New York Times*, March 20, 2002, A26.

133 "*In 1984 Weintraub published an article in the* Archives of Internal Medicine *promoting fen-phen as a powerful weight loss agent*": See M. Weintraub, J. D. Hasday, A. I. Mushlin, and D. H. Lockwood, "A Double Blind Clinical Trial in Weight Control: Use of Fenfluramine and Phentermine Alone and in Combination," *Arch Intern Med* 144 (6) (June 1986): 1143–1148.

"*The study was rejected by several journals before being published in 1992 in the* Journal of Clinical Pharmacology and Therapeutics": See M. Weintraub, "Long-term Weight Control: The National Heart, Lung, and Blood Institute Funded Multimodal Intervention Study," *J Clinical Pharm Ther* 51 (5) (1992): 581–646.

"*Everything must be known about them*": See Gina Kolata, "How Fen-Phen, A Diet 'Miracle' Rose and Fell," *New York Times*, September 23, 1997, F1. Kolata offers a penetrating account of the controversy in a series of *New York Times* reports.

"*There had been troubling reports linking fenfluramine, brand name Pondimin, with pulmonary hypertension, a rare but dangerous lung condition*": See F. Brenot et al., "Primary Pulmonary Hypertension and Fenfluramine Use," *British Heart Journal* 70 (6) (1993): 537–541.

134 "*For-profit weight loss centers, like Jenny Craig and Nutri/System, incorporated fen-phen into their programs*": Robert Kushner, "The Treatment of Obesity: A Call for Prudence and Professionalism," *Arch Internal Medicine* 157 (March 24, 1997): 602–604.

"*Wyeth-Ayerst Laboratories, a subsidiary of AHP that distributed the drug, could barely keep up with the demand*": See Jon Marcus and Sean Flynn, "Bitter Pill," *Boston Magazine* (December 1999): 108–161.

135 "*George Bray was one of several obesity experts speaking on the petition's behalf*": From the official transcript, open session, FDA Endocrinologic and Metabolic Drugs Advisory Committee, Thursday, September 28, 1995.

"*that both Bray and Manson regarded obesity as a deadly evil*": Dr. Manson had just that month published a controversial article in the *NEJM* that offered evidence that even a bit too much weight could be deadly: J. E. Manson, W. C. Willett, M. J. Stampfer, G. A. Colditz, D. J. Hunter, S. E. Hankinson, C. H. Hennekens, and F. E. Speizer, "Body Weight and Mortality Among Women," *New England Journal of Medicine* 333 (11) (September 16, 1995): 677–685. The article recorded as having the lowest mortality rate "women who weighed at least 15 percent less than the U.S. average for women of similar age and among those whose weight had been stable since early adulthood."

"*They spoke of animal studies that had implicated dexfenfluramine as a neurotoxin that might disrupt the structure and function of certain brain structures*":

See G. A. Ricarte, M. E. Molliver, M. B. Martello, J. L. Katz, M. A. Wilson, and A. L. Martello, "Dexfenfluramine Neurotoxicity in Brains of Non-Human Primates," *Lancet* 338 (1991): 1487–1488. U. McCann, G. Hatzidimitriou, A. Ridenour, C. Fischer, J. Yuan, J. Katz, and G. Ricurte, "Dexfenfluramine and Serotonin Neurotoxicity: Further Preclincal Evidence that Clinical Caution is Indicated," *J Pharmacol Experimental Therapeutics* 269 (1994): 792–798.

136 "*further delay the launch of what some analysts estimated would be a $1.8 billion drug*": David Willman, "Redux: Unheeded Warnings on Lethal Diet Pill," *Los Angeles Times,* December 20, 2000. Willman won the Pulitzer Prize the following year for his pioneering exposé of seven unsafe prescription drugs (including Redux) that had been approved by the Food and Drug Administration, and an analysis of the policy reforms that had reduced the agency's effectiveness.

"*Nonetheless, the testimony was, if anything,* more *damning of Redux than it had been in the first hearing*": From the official transcript, open session, FDA Endocrinologic and Metabolic Drugs Advisory Committee, Thursday, November 16, 1995.

137 "*People had uneasy feelings, I think, on both sides*": See Gina Kolota (1997), op. cit.

"*And thanks to strenuous efforts on the part of American Home Products and its representatives, it went out without the black box warning*": Internal Wyeth memo dated November 22, 1995, from J. Bathish to Fred Hassan re: Dexfenfluramine Assessment. Hassan wanted "a worst-case assessment" of the approval of Redux in the U.S. as it relates to labeling and the CPMP action for December 1995. "While we have a very difficult and arduous task of negotiating labeling and Phase IV commitments with the Agency over the next few weeks, every attempt will be made to ensure that no 'Black Box' Warnings, restrictions of use or negative statements find their way into the Redux labeling. . . . We will make every effort to neutralize these initiatives."

138 "*Two of the articles were subsequently published in peer-reviewed publications owned by Excerpta Medical Journals*": The story of the ghostwritten articles minimizing the risks and extolling the benefits of Redux , which first appeared in the *Dallas Morning News,* was recounted by Karen Birmingham, "Lawsuit Reveals Academic Conflict of Interest," *Nature Medicine* 5 (7) (July 1999): 117.

"*extolled the weight loss benefits of Redux*": Albert Stunkard, "Current Views on Obesity," *Am J Medicine* 100 (2) (February 1990): 230–236.

"*generally associated with a decrease in risk factors and the alleviation of clinical symptoms*": See FX Pi-Sunyer, "A Review of Long-Term Studies Evaluating the Efficacy of Weight Loss in Ameliorating Disorders Associated with Obesity," *Clin Ther* 18(6) (November-December 1996): 1006–1035; discussion 1005.

138 "*The FDA responded with a Public Health Advisory*": See FDA Public Health Advisory, July 8, 1997.

"*the FDA yanked fenfluramine and dexfenfluramine from the market*": See FDA press release, September 15, 1997.

"*But for hundreds, perhaps thousands, of people, it was already too late*": see Gina Kolata, "2 Top Diet Drugs Are Recalled Amid Reports of Heart Defects," *New York Times*, 1, September 16, 1997.

139 "*From the early 1960s, it was widely prescribed as a diet aid*": C. K. Haddock, W. S. Poston, P. L. Dill, J. P. Foreyt, and M. Ericsson, "Pharmacotherapy for Obesity: A Quantitative Analysis of Four Decades of Published Randomized Clinical Trials," *Int J Obes Relat Metab Disord* 26 (2) (February 2002): 262–273.

"*sleeplessness, anxiety, and, in some cases, psychosis closely resembling paranoid schizophrenia*": The scientific literature on amphetamines is immense, and the incidence of side effects common. For an early report of amphetamine-induced psychosis see, for example, P. J. Perry and R. P. Juhl, "Amphetamine psychosis," *Am J Hosp Pharm* 34 (8) (August 1997): 883–885.

141 "*company announced a dazzling 17 percent leap in operating profits for the previous quarter*": Matthew Herper, "Disaster of the Day: American Home Products," Forbes.com, January 20, 2001. forbes.com/2001/01/26/0126 disaster.html.

"*The pharmaceutical industry is by far the most profitable in the United States*": See, for example, E. Tanouye, "Drug Dependency: U.S. Has Developed an Expensive Habit, Now, How to Pay for It?" *Wall Street Journal*, November 16, 1998, and "How the Industries Stack Up," *Fortune*, April 17, 2000.

142 "*sales that year of $146 million, translating into 1.5 million dispensed prescriptions*": See Carolyn Wilhelm, "Growing Market for Anti-Obesity Drugs," *Chemical Market Reporter*, New York, May 15, 2000.

"*The following year U.S. sales soared to more than $207 million*": Interview with John LaRosa.

"*Doctors feel better about prescribing Xenical, and patients feel better about taking it*": Charles Billington, former president of the North American Association for the Study of Obesity (NAASO) and a professor of medicine at the VA Medical Center in Minneapolis, was kind enough to give me an overview of current views on drug treatments for obesity in a series of interviews in October 2000.

"*One obesity researcher likens Xenical to 'gastric Antabuse'*": Thomas Wadden of the University of Pennsylvania as quoted in Karen Springer, "Finally, the Free Lunch?" *Newsweek*, February 1, 1999.

143 "*meals containing more fat than recommended*": As stipulated in advertisements for Xenical and in the product warnings issued by maker Roche Pharmaceuticals.

"*averaging about six and one-half pounds more weight loss than experienced with a placebo*": See S. O'Meara, R. Riemsma, L. Shirran, L. Mather, and G. ter Riet, "A Rapid and Systematic Review of the Clinical Effectiveness and Cost-Effectiveness of Orlistat in the Management of Obesity," *Health Technol Assess* 5 (18) (2001): 1–81.

"*when xenical starts to lose its effect*": For an excellent review of Orlistat see John Garrow, "Flushing Away Fat," *British Medical Journal* 317 (1998): 830–831.

"*including adverse complications for diabetics*": See S. T. Azar and M. S. Zantout, "Diabetic Ketoacidosis Associated with Orlistat Treatment," *Diabetes Care* 24 (March 2001): 602.

"*and liver failure*": J. L. Montero, J. Muntané, E. Fraga, et al., "Orlistat Associated Subacute Hepatic Failure," *J Hepatol* 34 (January 2001): 173.

144 "*most of them promising weight loss or, to a lesser degree, an energy boost*": The FDA had until recently no jurisdiction over dietary supplements, herbal medicines, vitamins, hormones, or homeopathic remedies. But after thirty people died unexpectedly while taking l-tryptophan, a "natural" sleep aid, Congress passed the 1994 Dietary Supplement Health and Education Act, supposedly giving the FDA the power to restrict the distribution of herbal medications when the public health is at risk. This law had the paradoxical effect of making it even more difficult to regulate supplements by placing the burden of proof of harm on the government. Anyone familiar with the history of tobacco legislation knows how difficult it is to establish categorical evidence of harm.

"*concern about the risks of these products, given that they have no scientifically established benefits*": See Christine Haller et al., "Adverse Cardiovascular and Central Nervous System Events Associated with Dietary Supplements Containing Ephedra Alkaloids," *New England Journal of Medicine* 343 (25) (December 26, 2000): 1833–1838.

"*including eighty-one deaths, between January 1993 and February 2001*": See Philip J. Hilts, "Petition Urges U.S. to Ban Supplements With Ephedra," *New York Times*, September 6, 2001.

145 "*former California police officer, chauffeur, and real estate agent was implicated in a methamphetamine lab bust*": George Cowley and J. Reno, "Mad About Metabolife," *Newsweek*, October 1999, 51–55; also see earlier reports by Yvonne Abraham in *Boston Globe*, May 16, 1999, B8; and Charles R. Babcock in *Washington Post*, May 24, 1999, A1.

146 "*preparations of ephedrine and caffeine are safe when used according to the directions*": See FDA public meeting on The Safety of Dietary Supplements

Containing Ephedrine Alkaloids, August 8 and 9, 2000, www.fda.gov/ohrms/dockets/dockets/oon1200/tr00001b.rtf.

149 "*to eat anything and everything that they want, but simply burn it off as heat*": For my understanding of uncoupling proteins I owe a great debt to Leslie Kozak, one of the leading experts in this field. Dr. Kozak was kind enough to spend an entire afternoon discussing this complex and fascinating topic with me in his office at Pennington Biomedical Research Center in Baton Rouge.

"*free radicals, the nasty chemical rouges thought to hasten aging*": Harvard University researcher and physician Eleftheria Maratos-Flier was kind enough to pose the issue of free radicals in a discussion we had about uncoupling proteins and their potential role in weight loss.

"*But given the growing body of research linking longevity with low calorie consumption*": Laboratory and clinical evidence points to caloric restriction as a way to provide longevity and good health in nonhuman primates; see, for example, E. J. Masoro, "Caloric Restriction and Aging: an Update," *Experimental Gerontology* 35 (2000): 299–305, and A. S. Nicolas et al., "Caloric Restriction and Aging," *Journal of Nutrition, Health, and Aging* 3(2) (1999): 77–83. Studies in humans are so far lacking, but calorie restriction does show promise in terms of its benefit in reducing chronic disease symptoms. See, for example, Pekkarinen et al., "Weight Loss with Very-Low-Calorie Diet and Cardiovascular Risk Factors in Moderately Obese Women: One-Year Follow-Up Study Including Ambulatory Blood Pressure Monitoring," *International Journal of Obesity and Related Metabolic Disorders* 22 (1998): 661–666.

CHAPTER 8
SPAMMED

151 "*The Life of Samuel Johnson*": See James Boswell, *The Life of Samuel Johnson*, Vol. 3 (1791) (New York: Heritage Press, 1963), 294–295.

153 "*giraffe of vegetables*": See Robert Lewis Stevenson, *In the South Seas: Being an Account of Experiences and Observations in the Marquesas, Paumotus and Gilbert Islands* (University of Hawaii Press, 1971).

160 "*diabetes mellitus global epidemic is just the tip of a massive social problem now facing developing countries*": See Paul Zimmet and Pierre Lefebre, "The Global NIDDM Epidemic: Treating the Disease and Ignoring the Symptom," *Diabetologia* 39 (1996): 1247–1248.

"*emigrants to Fiji, South Africa, and Britain; and Chinese emigrants in Singapore, Taiwan, and Hong Kong*": See, for example, B. Joffee and P. Zimmet, "The Thrifty Genotype in Type 2 Diabetes: An Unfinished Symphony Moving to its Finale?" *Endocrine* 9 (2) (October 1998): 139–141.

161 "*black eyes full of fire and a mouth full of superb teeth . . . but a tendency to become fat*": See Harvey Gordon Segal, *Kosrae: The Sleeping Lady Awakens,* Kosrae State Tourist Division, Federated States of Micronesia (1989), 49. Segal, whose book is the definitive history of Kosrae, was kind enough to spend the better part of an afternoon teaching me about the history and culture of his adopted island. Segal has lived in Kosrae for decades and teaches college there, but spent the first part of his adult life teaching high school science in Newton, Massachusetts.

162 "*drink too much kava, and are too indolent. Sydney vessels brought the pox to the island*": Harvey Gordon Segal recorded these diary entries in *Kosrae: The Sleeping Lady Awakens,* but since he does not make use of footnotes, the original source is unclear.

"*to the hormone insulin in order to insure adequate blood glucose levels in the brain during periods of famine*": See James V. Neel "Diabetes mellitus: A 'Thrifty' Genotype Rendered Detrimental by 'Progress'?" *Am J Hum Genet* 14 (1962): 353–352.

163 "*Societies everywhere have, during some periods in history, lived under conditions of 'food stress'* ": See, for example, John S. Allen and Susan M. Cheer, "Civilization and the Thrift Genotype," *Asia Pacific Journal of Clinical Nutrition* 4 (4) (1995): 341–342.

"*evolution was punctuated by a number of particularly harrowing bottleneck events have developed the most effective versions*": See, for example, Ben Campbell et al., "Diabetes: Energetics, Development and Human Evolution," *Medical Hypothesis* 57 (1) (2001): 64–69.

"*European visitors are, in Kosraeans,* protective *against obesity and diabetes*": Interviews with several members of Jeffery Friedman's team enhanced my understanding of their work in Kosrae. Of particular help (and patience) was Maude Blundell, a genetic counselor who has done a good deal of work on the island. Marcus Stoffel, a diabetologist and professor at Rockefeller University participating in the Kosrae effort, was also most generous with his time, on this and other questions.

164 "*gradually eliminated, bringing [diabetes] to its present [relatively] low frequency*": Jared M. Diamond, "Diabetes running wild," *Nature* 357 (June 4, 1992): 362.

"*But we now know that the overweight generally have a faster than average metabolic rate*": See, for example, C. S. Fox et al., "Is a Low Leptin Concentration, A Low Resting Metabolic Rate, Or Both the Expression of the 'Thrifty Genotype'? Results from Mexican Pima Indians," *Am J Clin Nutr* 68 (5) (November 1998): 1053–1057.

166 "*average French person was just a bit shorter than the average American, but weighed about fifteen pounds less*": Peter N. Stearns, *Fat History: Bodies and Beauty in the Modern West* (New York: New York University Press, 1997).

167 "*by any standards a privileged country*": See Charles Mann, "1491," *Atlantic Monthly*, March 2002, 41–54. In this fascinating and provocative article on civilization in the "New World" prior to the invasion of the European explorer, Mann both quotes Frenand Braudel and makes mention of the French famines.

"*slimmer by far than Americans, nearly one in ten French adults today is overweight*": See A. Basdevant, "Obesity and Public Health," *Ann Endocrinol* (Paris) 61 Suppl 6 (December 2000): 6–11.

"*IHD is borne by the developing world, and that percentage is expected to grow*": Stephanie Ounpuu et al., "The Global Burden of Cardiovascular Disease," *Medscape Cardiology* (2000).

170 "*majority of the world's population, but also culturally and climatically ripe for significantly increased soft drink consumption*": In the 1992 Coca-Cola Company's Annual Report, "World of Opportunity," as cited in Benjamin R. Barber, *Jihad vs. McWorld* (New York: Ballantine Books, 1995), 69.

"*a* faster *metabolic rate than did normal-weight people*": See D. A. Schoeller "Measurement of Energy Expenditure in Free Living Humans By Using Doubly Labeled Water," *J Nutr* 118 (1988): 1278–1289.

Chapter 9
The Child is Father of the Man

173 "*the good is [having] enough to eat and the rest is talk*": Bertrand Russell, *A History of Western Philosophy*, Part II, Chapter 23 (1945): 747.

"*from September 17, 1944, until liberation on May 5, 1945, famine descended on Holland*": Louis de Jong, *The Netherlands and Nazi Germany*, the Erasmus Lectures (Cambridge: Harvard University Press, 1990).

174 "*foundations of slag flecked with bits of coal were, as one observer wrote, 'turned into mines'*": Henri A. van der Zee, *The Hunger Winter* (Lincoln: University of Nebraska Press, 1998), 85.

"*starving man of his fellows, They only consisted of a stomach and certain instincts*": Ibid., 64.

"*Holland was seven times higher than the wartime norm*": Nicky Hart, "Famine, Maternal Nutrition and Infant Mortality," *Population Studies* 47 (1993): 27–46.

"*The newborns were, on average, smaller than normal, but those who survived showed no excess of obvious deformity*": See Clement A. Smith, "Effects of Maternal Undernutrition Upon the Newborn Infant in Holland (1944–45)," *Journal of Pediatrics* 30 (3) (1947): 229–243.

175 "*note higher rates of heart disease, diabetes, and other chronic disease and even mental illness*": See, for example, Alan S. Brown et al., "Further Evidence of Relation Between Prenatal Famine and Major Affective Disorder," *American Journal of Psychiatry* 2 (February 2000): 157, and Ezra Susser et al., "Schizophrenia after Prenatal Exposure to the Dutch Hunger Winter of 1944/45," *Archives of General Psychiatry* 49 (1992): 983–988.

"*These findings have been controversial*": Publications in the arena of fetal programming effects on adult health have soared since the late 1990s. See, for example, M. Desai et al., "Role of Fetal and Infant Growth in Programming Metabolism Later in Life," *Biological Reviews* 72 (1997): 329–348; C. M. Law et al., "Is Blood Pressure Inversely Related to Birth Weight?" *Journal of Hypertension* 14 (1996): 935–941; J. G. Eriksson et al., "Catch-Up Growth in Childhood and Death from Coronary Heart Disease: Longitudinal Study," *British Medical Journal* 318 (February 13, 1999); Daniel Lackland et al., "Low Birth Weights Contribute to the High Rates of Early-Onset Chronic Renal Failure in the Southeastern United States," *Archives of Internal Medicine* 160 (May 22, 2000); D. A. Leon et al., "Gestational Age and Growth Rate of Fetal Mass Are Inversely Associated with Systolic Blood Pressure in Young Adults: An Epidemiologic Study of 165,136 Swedish Men Aged 18 Years," *American Journal of Epidemiology* 152 (June 26, 2000): 597–604.

"*hotly disputed*": Barker's theory has attracted a growing number of admirers, but also some detractors. See, for example, Michael S. Kramer, "Association between Restricted Fetal Growth and Adult Chronic Disease: Is it Causal? Is it Important?" *American Journal of Epidemiology* 152 (7) (June 26, 2000): 605–608.

179 "*stunting by malnutrition early in life predicts obesity in adults*": Daniel J. Hoffman et al., "Energy Expenditure of Stunted and Non-Stunted Boys and Girls Living in the Shantytowns of São Paulo, Brazil," *Obesity Research* 8 Suppl 1 (October 2000): 29S.

"*generations in a feed-forward, upward spiral of increasing body weight across generations*": See, for example, Barry E. Levin, "The Obesity Epidemic: Metabolic Imprinting on genetically susceptible neural circuits," *Obesity Research* 8 (4) (July 2000): 342–347.

181 "*It is of national importance that the life of every infant be vigorously conserved*": As quoted without citation in Caroline Fall, "Hartfordshire's Babies," Medical Research Council Environmental Epidemiology Unit, Southampton.

185 "*permanent adaptation of central regulatory mechanisms of energy intake and expenditure*": See Anita C. J. Ravelli et al., "Obesity at the Age of 50 in Men and Women Exposed to Famine Prenatally," *Am J Clin Nutrition* 70 (1999): 811–816.

"*an obese fifteen-year-old has about an 80 percent chance of being obese as an adult*": Interview with Susan Roberts, professor of nutrition, Tufts University Medical School.

"*in 1999, the last year for which survey information is available*": CDC statistics available on website at www.cdc.gov

"*In Australia, childhood obesity rates tripled between 1985 and 1995, and today one out of every five children there is overweight*": See Anathea Magarey et al., "Prevalence of Overweight and Obesity in Australian Children and Adolescents: Reassessment of 1985 and 1995 Data Against New Standard International Definitions," *Medical Journal of Australia* 174 (11) (June 4, 2001): 561–564.

"*children and adolescents in Canada, Japan, India, Hong Kong, Singapore, Bangladesh, Libya, the United Kingdom, and New Zealand, among other countries*": A. Fagot-Campagna et al., "Type 2 Diabetes in Children: Exemplifies the Growing Problem of Chronic Diseases," *British Medical Journal* 322 (2001): 377–378.

186 "*risks that can haunt them into adulthood*": Telephone interview with Myles Faith, pediatric obesity specialist and associate research scientist in the obesity research center at St. Lukes Roosevelt Hospital, New York.

"*These are all precursors to adult heart disease*": Interview with Aviva Must, associate professor in the Department of Family Medicine and Community Health at Tufts University, who outlined the dangers on adult health of pediatric obesity and overweight, as did Susan Roberts.

"*collaborated on a landmark study of obesity and television viewing*": William Dietz et al., "Do We Fatten Our Children at the Television Set? Obesity and Television Viewing in Children and Adolescents," *Pediatrics* 75 (5) (1985): 807–812.

"*children watching four or more hours were obese*": See C. J. Crespo et al., "Television Watching, Energy Intake, and Obesity in U.S. Children: Results from the Third National Health and Nutrition Examination Survey, 1988–1994," *Archives of Pediatrics and Adolescent Medicine* 155 (3) (2001): 360–365.

"*children who watched five or more hours a day were more than eight times as likely to be overweight as those watching two hours or less*": Steven Gortmaker et al., "Television Viewing as a Cause of Increasing Obesity Among Children in the United States, 1986–1990," *Archives of Pediatric Adolescent Medicine* 150 (April 1996): 356–362.

"*The more television children watch, the more they eat*": C. J. Crespo, E. Smit, and R. P. Troiano, *Archives of Pediatrics and Adolescent Medicine* (2001), op cit.

187 "*at least in studies that have been done with obese children, perhaps because it engages their minds a bit more emphatically*": See William Dietz et al., "*Effect of sedentary activities on metabolic rate,*" *Am J Clin Nutr* 59 (1994): 556–559; and R. C. Klesges et al., "Effects of Television on Metabolic Rate: Potential Implications for Childhood Obesity," *Pediatrics* 91 (2) (1993): 281–286.

188 "*Goncharov wrote that 'physical education as a whole promotes the development of those qualities which are essential to future warriors of the Red Army'*": B. P. Yesipov and N. K. Gonchorov, *I Want to Be Like Stalin,* translated by George S. Counts and Nucia P. Lodge (New York: The John Day Company, 1947), 7.

"*concluded that American children, though among the best fed and healthiest in the world, were woefully under-exercised*": Sonya Weber, "Kraus-Weber Tests," *Pennsylvania Journal of Health, Physical Education and Recreation* (September 1956): 14–15.

189 "*described the effort as 'primarily one of vaguely worded publicity and promotion releases for the fitness cause'*": "A Fit Week for a Second Look," *Sports Illustrated,* May 26, 1958, 37–39, as cited in C. W. Hackensmith, *History of Physical Education* (New York: Harper & Row, 1966).

"*Seaton opined publicly that the best way to become fit was to take up bird watching*": Ibid.

"*promotion of sports participation and physical fitness is a basic and continuing policy of the United States*": John F. Kennedy, "The Soft American," *Sports Illustrated* 13 (December 26, 1960): 17.

"*than half of high school students were enrolled in physical education classes for even one hour a week*": The Centers for Disease Control Fact Sheet on Youth Behavior Trends can be found at www.cdc.gov/od/oc/media/fact/youthrisk.htm.

190 "*In addition to the more readily modifiable factors, high crime level was significantly associated with a decrease in weekly moderate to vigorous physical activity*": See Penny Gordon-Larsen et al., "Determinants of Adolescent Physical Activity and Inactivity Patters," *Pediatrics* 105 (6) (June 2000).

Chapter 10
An Arm's Reach from Desire

191 "*We may not pay Satan reverence, for that would be indiscreet, but we can at least respect his talents*": Mark Twain, "Concerning the Jews," in *Harper's* (New York, September 1899), in Charles Neider, ed., *The Complete Essays of Mark Twain* (New York: Doubleday, 1985).

192 "*In* Kids as Customers": See James U. McNeal, *Kids As Customers: A Handbook of Marketing to Children* (Lanham: Lexington Books, 1992).

193 "*In his most recent book*": See James U. McNeal, *The Kids Market: Myths and Realities* (Ithaca, N.Y.: Paramount Market Publishing, 1999).

193 "*But in this scenario, the harried woman tosses Lunchables*": Not to miss out on the super-size trend, Kraft recently came out with a new line called *LUNCHABLES Mega Packs.* On its promotional website, Kraft describes the Deep Dish Pepperoni Pizza selection as follows: "Two deep dish pizza crusts topped with tasty pizza sauce, zesty pepperoni, and shredded mozzarella cheese, 12 oz. can of LUNCHABLES cola, Reese's® peanut butter cup."

194 "'*parallel pantries' where parents maintain a separate stash of favorite foods for the little ones in order to minimize 'dinnertime battles'*": See Leslie Sharra, Carol Cronk, and Audrey Nelson, "Foods for Children: Here come the Millenials," www preparedFoods.com, May 2001, 31.

"*116,000 new 'food products' were introduced to the U.S. market between 1990 and 1998*": See Anthony E. Gallo, "Fewer food products introduced in last 3 years," *FoodReview* 22 (3) (1999): 27–29.

195 "*offers the feel of freedom while diminishing the range of options and the power to affect the larger world*": Benjamin Barber, *Jihad vs. McWorld* (New York: Ballantine Books, 1995), 220–221. Like so much of Barber's writing and thinking, a bitingly witty and perceptive critique.

"*The nearly $112 billion American fast food industry claims it would be delighted to offer healthy options, but that, sadly, consumers—fools that we are—just don't buy them*": See James Matoriur, "Awareness Campaign Needed," *Nation's Restaurant News* (August 27, 2000): 32.

196 "*Overall sales slipped, and stockholders pressured the company to drop the low-fat line, which Taco Bell's did with little fanfare*": Ann Stone, "Lean? No Thanks," *Restaurants and Institutions,* May 15, 1997.

"*In blind taste, tests one hundred families agreed that the AU Lean trumped the fat burger for 'flavorfullness,' 'juiciness,' and 'likability'*": For a lively recounting of the McLean fiasco see Malcom Gladwell, "The Trouble with Fries," *New Yorker,* March 5, 2001.

198 "*We can't really address or defend nutrition. We don't sell nutrition, and people don't come to McDonald's for nutrition*": Internal company memo, March 1986, brought to light in the McLibel case, the longest court trial in British history, in which McDonald's sued two British Greenpeace activists for libel.

"*The Institute of Medicine Food and Nutrition Board is sustained in part by M&M Mars, and the American Society for Clinical Nutrition by BestFoods and Coca-Cola*": See Marion Nestle, *Food Politics: How the Food Industry Influences Nutrition and Health* (Berkeley: University of California Press, 2002), 113.

199 "*She looks to be in her early forties, and every inch the 'soccer mom with an attitude' type whom food marketers strive to cultivate*": While this conversation was technically on the record, I am keeping it anonymous to protect the source.

200 "*The time spent by an average customer in a fast food restaurant is a blistering eleven minutes*": E. Christine Jackson, "Ethnography of a Urban Burger King Franchise," *Journal of American Culture* 2 (3) (1979): 534–539, 537.

201 "*the latest year for which figures are available, dinner preparation had shrunk to fifteen minute*": Jane Snow, "Look at What They've Done with Frozen Pop-Up Foods." Reporter-news.com, June 21, 2000.

"'*no-think foods,' that by definition 'don't drip, crumble, require utensils, or demand inordinate attention*'": Comments of consumer expert Mona Doyle in Trish Hall, "Now Food for the Otherwise Engaged," *New York Times,* April 15, 1987.

"*Waiting has become an intolerable circumstance*": See Caroline Mayer, "Instant Everything: Making Fast Lane Faster," *Washington Post,* January 2, 2001.

"*The National Restaurant Association ranks in* Fortune *magazine's list of the twenty-five most powerful lobbies in Washington*": Fortune's 25 most powerful lobby list can be found on at www.fortune.com/lists/power25/index.html.

202 "*California's 7.25 percent sales tax seems to be a particular annoyance*": See, for example, Sheila R. Cohn, R. D., and Steven F. Grover, "Doing Battle with the 'Fat' Police," *Restaurant Hospitality,* December 2000, in which the authors state that "such regulations restrict customer choice and eliminate foods people like." What is interesting is that these same organizations lobby mightily for government intervention in the form of an increase from 50 percent to a 100 percent tax deduction for restaurant meals.

"*Eighty percent of that increase was captured by fast food outlets*": See NPD's Fourteenth Annual Report on Eating Patterns in America.

"*tends to give them higher turnovers per units than independents, thus boosting share of the market in terms of value sales*": See QSR, The Magazine of Quick Service Restaurant Success, www.qsrmagazine.com/issue/fertilesoil/part1–2.html.

"*McDonald's red and yellow ensign is the new version of America's star-spangled banner . . . whose cultural hegemony insidiously ruins alimentary behavior*": See Sophie Menunier, "The French Exception," *Foreign Affairs* (July/August 2000): 107.

203 "*In* Fast Food Nation, *journalist Eric Schlosser describes fast food as a revolutionary force that has transformed not only the American diet, but our landscape, economy, and popular culture*": See Eric Schlosser, *Fast Food Nation: The Dark Side of the All-American Meal* (Boston: Houghton Mifflin, 2001), a tightly argued, eye-opening invective.

204 "*In his collection of postmodern prose poems,* Letters to Wendy's": See Joe Wenderoth, *Letters to Wendy's* (Easthampton, Mass.: Vere Press, 2000).

"*McDonald's first Russian outpost, opened in 1990 near Moscow's Pushkin Square, remains today the world's busiest fast food outlet*": See Bill Keller, "Arise, Ye Prisoners of Starvation," *New York Times,* February 23, 2002, A31.

"*childhood obesity and overweight in Russia, once hardly a problem, rose to 16 percent*": See Youfa Wang, "Cross-National Comparison of Childhood Obesity: The Epidemic and the Relationship Between Obesity and Socioeconomic Status," *Int J Epidemiol* 30 (2001): 1129–1136.

205 "'*cheese fries,' chicken fingers, and nachos constitute 30 percent of restaurant profits*": See William A. Roberts, Jr., "An Appetizing Revolution," PreparedFoods.com, May 2001.

"*in the form of chips, French fries, and frozen hash browns than we did thirty years ago*": See, for example, G. H. Sullivan and L. D. Green, "Potatoes: Production and Marketing Trends," Purdue University Dept. of Horticulture Agricultural Exp. Station Bulletin, 1990, in Edwin Plissey, "Potato Production Utilization Trends in North America," prepared for the University of Maine, 1998.

206 "*Indeed, cheese has become an imperative, the common denominator in our melting pot cuisine*": Judy Putnam, an economist with the Food and Rural Economics Division of the USDA Economic Research service, wrote in a report in *FoodReview* (Vol. 22, Issue 3, 2–12) that in 1998, Americans drank on average 35 percent less milk and ate nearly four times as much cheese (excluding cottage cheese) than we did in the 1950s. We now eat each year about 28.4 pounds of cheese per person, two-thirds of it in commercially prepared and manufactured foods.

"*single-handedly reversing a collapse in pork prices*": See Michael Pearson, "Lower Production, Rising Demand for Fast-Food Bacon Restores Profitability to Hog Farming," *Associated Press,* April 21, 2000.

207 "*one recaptures infantile experiences, regressions and transgressions*": See Claude Fischler, "La 'Macdonaldisation' des moeurs," in J. L. Flandrin and Massimo Montanari, eds., *Historire de l'alimentation* (Paris: Fayard, 1966), 859–879, as translated by cultural anthropologist Sidney W. Mintz in notes on his essay, "Swallowing Modernity," in James L. Watson, ed., *Golden Arches East* (Stanford University Press, 1997), 236.

"*Chinese and Korean men in particular report that they find the food* "*unfilling*" *and* "*less than satisfying*": See Sangmee Bak, "McDonald's in Seoul," ibid., 137–160.

210 "*thirty-four teaspoons daily per person by the most recent USDA estimates, a nearly 30 percent increase from only fifteen years ago*": See Judy Putnam

et al., "Per Capita Food Supply Trends: Progress Toward Dietary Guidelines," *FoodReview* 23 (3) (September-December 2000).

"*has jumped by thirty-two pounds per person since 1970*": see Judith Jones Putnam and Jane E. Allshouse, Food Consumption, Prices, and Expenditures, 1970–97, Food and Rural Economics Division, Economic Research Service, U.S. Department of Agriculture, Statistical Bulletin No. 965.

211 "The Lancet *that at least for overweight children, soft drink consumption directly predicts weight gain*": See David Ludwig et al., "Relation between consumption of sugar-sweetened drinks and childhood obesity: a prospective, observational analysis," *Lancet* 357 (February 17, 2001): 505–508.

212 "*test-market, promote sampling and trial usage and—above all—to generate immediate sales*": See John Sheehan, "Why I Said No to Coca-Cola," *Rethinking Schools* 14 (2) (winter 1999).

"*soft drinks within arm's reach of desire . . . schools are one channel we want to make them available in*": See R. Pear, "Senator, Promoting Nutrition, Battles Coca-Cola," *New York Times,* April 26, 1994, A20.

"*Congress to allow school sales of soda and other snack foods over the continued objections of the USDA*": For an excellent discussion of the selling of soft drinks in schools, see Marion Nestle, "Soft Drink Pouring Rights: Marketing Empty Calories," *Public Health Reports* 115 (2000): 308–319.

"*This sort of deal is commonplace—roughly one-sixth of school districts*": See Derrick Z. Jackson, "The Other Epidemic: Deadly Obesity," *Boston Globe,* December 14, 2001, A27.

213 "*some school districts joined forces with the National Soft Drink Association to thwart the effort in federal appeals court*": See Greg Winter, "States Attempt to Curb Sales of Soda and Candy in Schools," *New York Times,* September 9, 2001, A1.

"'*Big Brother' in the form of government injecting itself into decisions when it comes to refreshment choices*": See U.S. Senate Report 103–300, "Better Nutrition and Health For Children Act of 1994. 103rd Congress, 2nd. Session," July 1, 1994.

"*leading coffee makers to position coffee as a cold, sweet drink, much like soda*": See NPD Group 14th Annual Report on Eating Trends.

"*sweetened desserts than do Western children, and until quite recently consumed less ketchup, jam and other sweet condiments*": Many thanks for fascinating insights on Japanese cuisine and tastes to my friend June Kinoshita, author of the indispensable *Gateway to Japan,* 3rd edition (Tokyo: Kodansha International, 1998), widely agreed to be the world's finest guidebook to Japan.

214 "*Soft drinks are threatening to overtake the tea culture in India and Indonesia*": Benjamin R. Barber, *Jihad vs. McWorld* (New York: Ballantine Books, 1995), 68–72.

"*from a can or bottle as though blowing on a trumpet* (*rappa*)—*at least for the young*": See Ohnuki-Tierney, "McDonald's in Japan: Changing Manners and Etiquette," in James L. Watson, ed. (1997), op. cit., 161–182.

"*there's a good chance that this could be part of our problem*": Corn production is at an all-time high at the same time that new technologies have lowered the cost of producing high-fructose corn syrup. According to the USDA Economic Research Service, United States production of HFCS jumped from 327 million bushels in 1985 to 550 million bushels in 2000.

215 "*that consuming a diet that is more than 30 percent fat elevates the desire for fat and carbohydrates on a physiological level*": Based on interviews with Dr. Leibowitz, and see also Sarah F. Leibowitz and Bartley B. Hoebel, "Behavioral Neuroscience of Obesity," 313–358, in George Bray, Claude Bouchard, and W. P. T. James, eds., *Handbook of Obesity* (New York: Marcel Dekker, 1998).

216 "*decade one thousand new low-fat and reduced-fat products swamped the market, yet fat consumption has decreased—if at all—only slightly*": See Laura S. Sims, *The Politics of Fat: Food and Nutrition Policy in America* (Armonk, N.Y.: M. E. Sharpe, 1998). Also, thanks to Larry Linder, executive editor of the Tufts Health and Nutrition Letter, for our conversations on the food industry's efforts to capitalize on the nation's growing fear of fat.

"*40 percent of total calories thirty years ago, to about 34 percent of total calories*": See Walter Willett, *Eat Drink and Be Healthy* (New York: Simon and Schuster Source, 2001).

217 "*overfed a thousand calories a day for two months*": See James A. Levine et al., "Role of Nonexercise Activity Thermogenesis in Resistance to Fat Gain in Humans," *Science* (January 8, 1999): 212–214.

"*determining individual differences in response to overeating, at least over the short term*": See, for example, S. Snitker et al., "Spontaneous Physical Activitity in a Respiratory Chamber Is Correlated to Habitual Physical Activity," *International Journal of Obesity* 25 (10) (October 2001): 1481–1486; Tamaki Matsumoto, Chiemi Miyawaki, Hidetoshi Ue, Tomo Kanda, Yasuhide Yoshitake, and Toshio Moritan, "Comparison of Thermogenic Sympathetic Response to Food Intake between Obese and Non-obese Young Women," *Obesity Research* 9 (2001): 78–85.

218 "*15 to 20 percent of calories eaten, while storing dietary fat as body fat requires only 3 percent of calories consumed*": Interview with Claude Bouchard, Executive Director, Pennington Biomedical Research Center, Louisiana State University, Baton Rouge.

CHAPTER 11
THE RIGHT CHOICE

220 "*great literature, great art, great music, a functioning democracy*": Benjamin R. Barber in an interview with *New Perspectives Quarterly,* September 22, 1995.

221 "*an enzyme that seems to encourage fat storage in mice, is, as I write, undergoing human trials in Britain*": See press release issued by Millennium Pharmaceuticals, "Millennium and Abbott Initiate Phase I Trials of Their First Genomically-Derived Drug Target for Obesity," November 27, 2001.

222 "*This 'law of constant travel times' has deep roots*": See, for example, G. Hupkes, "The Law of Constant Travel Time and Trip Rates," *Futures* 14 (1) (1988): 38–46.

223 "*its towering canopy of ancient oaks, poplars, dogwoods, maples, and magnolias*": See Tom Watson, "City Trees," *The Journal of The Society of Municipal Arborists* 36 (6) (November/December 2000).

224 "*In Atlanta that adds up to fifty-three hours a year—a full working week—sitting and waiting for traffic to clear*": See "Trends, Implications and Strategies for Balanced Growth in the Atlanta Region," prepared by the SMARTRAQ research program at the Georgia Institute of Technology. For an interesting comparison see R. Michael et al., *Sex in America: A Definitive Survey* (Boston: Little, Brown, 1994). While Americans claim sex as their favorite activity, this survey of 3,432 adults aged eighteen to sixty gives evidence that we actually engage in it on average a mere one-half hour per week, which is less than half the time we average sitting in traffic jams.

225 "*uncovered a connection between an elevation of enzyme activity regulating stress hormones—such as cortisol—and overeating and obesity in mice.*": Hiroaki Masuzaki, Janice Paterson, Hiroshi Shinyama, Nicholas M. Morton, John J. Mullins, Jonathan R. Seckl, and Jeffrey S. Flier, "A Transgenic Model of Visceral Obesity and the Metabolic Syndrome," *Science* (December 7, 2001): 2166–2170.

"*Endocrinologists at the University of Edinburgh had earlier found an elevated level of cortisol in the fat tissue of obese men*": See Eva Rask, Tommy Olsson, Stefan Soderberg, Ruth Andrew, Dawn Ew Livingston, Owe Johnson, and Brian Walker, "Tissue-Specific Dysregulation of Cortisol Metabolism in Human Obesity," *Journal of Clinical Endocrinology and Metabolism* 86 (3) (March 2001).

226 "*American men now average about forty hours of free time per week, and American women average thirty-nine*": See John P. Robinson, Geoffrey Godbey, and Anne Jaap Jacobson, *Time for Life: The Surprising Ways Americans Use Their Time,* 2nd edition (Pennsylvania State University Press, 1999).

227 "*advertising on children's television programs is least regulated—in Australia, the United States, and England*": See, for example, J. K. Binkley, J. Eales, and

M. Jekanowski, "The Relation Between Dietary Change and Rising U.S. Obesity," *International Journal of Obesity* (24) (2000): 1032–1039, and C. J. Crespo, E. Smit, and R. P. Troiano, "Television Watching, Energy Intake and Obesity in U.S. Children: Results from the Third National Health and Nutrition Examination Survey, 1988–1994," *Archives of Pediatrics and Adolescent Medicine* 155 (3) (2001): 360–365.

227 "*350,000 television commercials before they reach voting age*": Center for Media Education, "Marketing to Children Harmful: Experts Urge Candidates to Lead Nation in Setting Limits," an open letter sent to presidential candidates on October 18, 2000.

"Teletubbies, *a public television program targeted to the preschool set, is sponsored by McDonald's*": The 1934 Communications Act flatly forbid noncommercial broadcasters from airing messages that "promote any service, facility or product" for profit, but donor acknowledgments that "identify but don't promote" are permitted. Thanks to this loophole Chuck E. Cheese pizza restaurants proudly sponsor children's programming on PBS, as do Tony the Tiger and Ronald McDonald.

228 "*pose restrictions on children's advertising, and are pressing the other states of the European Union to do the same*": See Jonathon Thompson, "Pressure for Ban on Ads During Children's TV," *The Independent,* February 17, 2002.

"*School districts in California, Texas, Washington state and elsewhere are actively fighting this trend with great success, and they are to be applauded for their efforts*": See Timothy Egan, "In Bid to Improve Nutrition, Schools Expel Soda and Chips," *New York Times,* May 26, 2002.

229 "*much of the Western world even before the widespread application of antibiotics and other curatives*": An interesting essay on the rise and fall of environmental linked disease is Jess H. Ausubel et al., "Death and the Human Environment: The United States in the 20th Century," *Technology in Society* 23 (2) (2000): 1131–1146. For my discussion of the impact of public health to effect disease I relied on Roy Porter's *The Greatest Benefit to Mankind* (New York: W. W. Norton, 1997).

230 "*Twenty-seven percent of Americans are already obese, double the level of just twenty years ago*": see Centers for Disease Control, National Center for Chronic Disease Prevention and Health Promotion, "Obesity and Overweight, a Public Health Epidemic," www.cdc.gov/nccdphp/dnpa/obesity/epidemic.htm. Obesity is defined as a Body Mass Index (BMI) of 30 kg/m2 or more.

"*twenty years of aging*": See Roland Sturm, "The Effects of Obesity, Smoking and Drinking on Medical Problems and Costs," *Health Affairs* 21 (2) (March/April 2002).

"*it did all in its formidable power to call into question the motives of advocates and agencies that threatened to expose it*": For the history and politics of the tobacco wars, I relied in particular on Michael Pertschuk, *Smoke in Their Eyes: Lessons in Movement Leadership from the Tobacco Wars* (Nashville: Vanderbilt University Press, 2001), and Richard Kluger, *Ashes to Ashes: America's Hundred-Year Cigarette War, the Public Health, and the Unabashed Triumph of Philip Morris* (New York: Vintage Books, 1997).

231 "*Obesity prevention, too, has been an explicit goal of U.S. public health policy since 1980*": See E. K. Battle and Kelly D. Brownell, "Confronting a Rising Tide of Eating Disorders and Obesity: Treatment vs. Prevention and Policy," *Addictive Behavior* 21 (1996): 755–765.

"*be an issue beyond partisan politics, and that cigarettes are a product that will never be tamed without government intervention*": Richard Kluger, "A Peace Plan for the Cigarette Wars," *New York Times Magazine,* April 7, 1996.

232 "*brand of candy bar—the wildly popular Milky Way—is $25 million*": See Stuart Elliott, "Mars and Other Marketers Offer Scaled-Down Spots Based on Simple Pleasures of Life," *New York Times,* April 1, 2002, C9.

235 "*Centers for Chronic Disease Prevention and Health Promotion*": Thanks to Marion Nestle of New York University for taking the time to share with me in a telephone interview her insights into the USDA and the American food industry.

"*Shiriki Kumanyika has pointed out 'it is the either/or nature of Christakis' observation that carries the insight'*": I was most fortunate to spend an afternoon discussing obesity and its public health implications with Shiriki K. Kumanyika, professor of epidemiology, Department of Biostatistics and Epidemiology at the University of Pennsylvania. Prof. Kumanyika's insights were remarkable in both their lucidity and their unyielding realism. For her thoughts on George Christakis see Shiriki K. Kumanyika, "Minisymposium on Obesity: Overview and Some Strategic Considerations," *Annual Review of Public Health* 22 (2001): 293–308.

Acknowledgments

Bill Whitworth, a man of great presence and uncanny modesty, has quietly helped legions of writers find their footing and voice, and years ago I was fortunate to be among them. Bill is an unfailingly courteous, gracious, and thoughtful man, and one of the increasingly rare editors who has faith enough in writers—and readers—to allow a story to speak for itself. As editor of the *Atlantic Monthly* he gave me a break—a big one—and I am forever grateful to him. Thanks also to *Atlantic Monthly* managing editor Cullen Murphy, and incisive senior editors Corby Kummer and Barbara Wallraff, for keeping me both cogent and honest. Current *Atlantic Monthly* editor Mike Kelly not only ran an excerpt of this book in the magazine, but suggested that I direct at least some attention to what he called the "marketing of obesity"—a brilliant stroke.

Over the several years I spent thinking about and working on this book, I received ideas and advice from scores of people. Dr. Jeffery Friedman opened his mind and heart to the project—offering crisp insights and lucid commentary at what for a time was on a weekly basis. A world-renowned scientist with a schedule to match, he somehow found time to patiently coach me through a daunting scientific labyrinth.

Dr. Rudy Leibel is a remarkable man, a humanist, a humorist, an inspiration to a generation of medical students, a top researcher, a compassionate physician, and the father of daughters who are themselves well on their way to becoming doctors. He knows much more than he shares, and he shares more than anyone should risk hoping for. Drs. Nathan Bahary, Don Seigel, and Yiying Zhang were forthright and patient in their personal accounts of the race to clone the obese gene, and Nathan in particular was deeply generous in his meticulous accounting of the discovery. Dr. Stephen O'Rahilly showed me an awfully good time in Cambridge, England, where we met to talk about single-gene mutations that cause obesity in humans.

In Cambridge, thanks also to Dr. Sadaf Farooqi, and to Stephen's wife, Suzie, for enduring that five-hour science fest that Steve had billed as a "social lunch." Thanks also to Dr. David Barker, who met with me in his institute and then in his home in the south of England to hash out his theory of prenatal programming, and to his wife, Jan, who is such gracious company.

No amount of gratitude would be enough to repay the kindness and generosity of the extraordinary medical epidemiologist Dr. Steven B. Auerbach, who at his own expense accompanied me for two weeks to the Micronesian islands of Kosrae and Pompei, serving as a guide, scientific interpreter, and friend.

Dr. David Allison, professor of biostatistics, University of Alabama at Birmingham, framed some of the questions raised by this book, and pointed me toward many of the answers. Dr. Charles Billington, former president of the North American Association for the Study of Obesity, offered invaluable guidance and advice, as well as read and commented on an early draft of the book. Dr. Albert Stunkard, professor of psychiatry at the University of Pennsylvania, spent hours with me recounting the early psychoanalytic theories.

In Boston I want to thank especially Drs. Jeffery Flier, Eleftheria Maratos-Flier, Joel Elmquist, Bradford Lowell, Susan Roberts, George Blackbrun, David Ludwig, Aviva Must, Rose Frisch, Stephen Gortmaker, Andrew Greenberg, Isaac Greenberg, Bruce Spiegelman, and Walter Willett. Dr. Edward Munn, a compassionate and gifted surgeon, granted me unfettered access to his surgical suite, and his patient Nancy Wright granted unfettered access to her life. Best of health to you, Nancy. At Millennium Pharmaceuticals Dr. Louis Tartaglia offered insights into the daunting challenge of designing antiobesity drugs, and Steve Holtzman got the ball rolling.

At Jackson Laboratories in Bar Harbor, Dr. Douglas Coleman squandered a day he might otherwise have spent outdoors in productive labor to speak with me. Doug, I sure do hope the story has been told straight this time. Also helpful at Jackson was Dr. Patsy Nashina and Joyce Peterson.

At the Environmental Epidemiology Unit at the University of Southampton I thank Drs. Sarah Duggleby, Mark Hanson, David Phillips, Caroline Fall, and Hazel Inskip. Also in Cambridge, U.K. I am deeply grateful to Dr. Andrew Prentice, who not only spent long hours speaking with me about his studies of food deprivation in the Gambia, but donated fresh batteries to rejuvenate my flagging tape recorder. At Gemini Research thanks to Dr. Patrick Kleyn, at the University of Leeds to Dr. John Blundell, and at Imperial College School of Medicine in London to Dr. Stephen Bloom.

At Rockefeller University I am indebted for their time to Drs. Marcus Stoffel, Jan Breslow, James Darnell, Stephen Burley, Sarah Leibowitz, and Mary Jeanne Kreek, and also to Maude Blundell, whose important human genetic studies in Kosrae I hope were not too much compromised by my inquires. At the Obesity Research Center at St. Luke's Hospital at Columbia University College of Physicians and Surgeons I thank Drs. Joseph Vasseli, Myles Faith, and Stephen Heymsfield.

In Baton Rouge, I thank Dr. Claude Bouchard, director of Pennington Biomedical Research Center, for our discussion of the history of the genetic underpinnings to obesity. Also at Pennington thanks to Drs. David York, Steven Smith, Leslie Kozak, and George Bray.

Thanks to Dr. Dale Schoeller for speaking to me several times about his landmark studies of doubly labeled water, and to Dr. Eric Ravussin for his illumination of the set point theory. Thanks to Drs. Randy Seeley and Michael Schwartz for getting me up to speed on the history of the search for a satiety hormone. Thanks to Dr. Arthur Campfield for making clear that the cloning of the *ob* gene involved a long-standing collaboration between peers, and to Dr. Adam Drewnoski, who explained why the choices we make in a "free society" are rarely free ones.

In Atlanta, Dr. William Dietz at the Centers for Disease Control and Prevention was marvelously available, despite a schedule made even more hectic by the horrifying events of September 11. Thanks as always, Bill. Also at the CDC, thanks to Dr. David Williamson and Dr. Richard Jackson. Many thanks also to Dr. Howard Frumkin, professor and chair of the Department of Environmental and Occupational Health at Emory University's Rollins School of Public Health, who made emphatically clear the connection between urban environment and chronic disease. Also thanks to Dr. Larry Frank at Georgia Institute of Technology, and to Dr. Shiriki K. Kumanyika who met with me in Atlanta, but whose professional home is the Center for Clinical Epidemiology and Biostatistics, University of Pennsylvania School of Medicine.

For enduring sometimes lengthy interviews I thank Dr. Greg Barsh of Stanford University; Dr. Claude Lenfant, director of the National Heart, Lung, and Blood Institute of the National Institutes of Health; Dr. Michael Stock of St. George's Hospital Medical School in London; Dr. Phillip James, chairman of the International Obesity Task Force; Dr. Paul Zimmet, professor and director of the International Diabetes Institute in Australia; Dr. Robert Eckel of the University of Colorado Health Sciences Center; Dr. Ernst Schafer of the Jean Mayer USDA Human Nutrition Research Center on Aging at Tufts University; and Dr. Barry Popkin, in the School of Public Health at the University of North Carolina at Chapel Hill. Thanks also to Dr. James Hill, professor of pediatrics and director of the Clinical Nutrition Research Unit at the University of Colorado Health Sciences Center, and Dr. Barbara Hansen, professor of physiology at the University of Maryland School of Medicine.

Dr. Bruce Schneider, who met with me in Washington, D.C., was extraordinarily generous—and brave—in expressing his concern over FDA approval of certain drugs used for obesity treatment. Dr. Cutberto Garza, professor and assistant provost at Cornell University and a world's expert on childhood nutrition, spoke with me for hours about the role of early life experience in predicting adult health.

Dr. Marion Nestle was willing to speak with me about the contribution of the food industry to the obesity crises, despite her own tight book deadline, as was Dr. Michael Jacobson, executive director of the Center for Science in the Public Interest.

At Boston University, many people sacrificed to make this book possible. Thanks to Dean Brent Baker and to special projects director Bill Ketter for granting me a much-needed teaching leave, and to journalism department chairman Bob Zelnick for backing me every step of the way. Friends and colleagues Caryl Rivers and Susan Blau were as always a wonderful support.

Thanks to my endlessly patient, exacting, and brilliant research assistant Dr. Robin Orwant, who is living proof that earning a Ph.D. in microbiology from Harvard doesn't mean you can't be a great journalist. Thanks also to my earlier research assistants, Paroma Basu, Jessica Penny, and Dr. Sylvia Pagen Westphal, great journalists all. Thanks also to the helpful staff in the rare books room at the Countway Library at the Harvard Medical School, who generously gave me access to one of the world's finest collections of medical history texts.

Eternal gratitude to dear friends John Horgan, Robin Marantz Henig, and Marcia Bartusiak, writers who took time away from their own important books-in-progress to read and comment on a draft of mine. Special thanks also to Ellie McCarthy, a gifted editor whose keen eye has not dimmed with the demands of motherhood.

As happens with so many books in these turbulent times, this project was acquired by one editor and edited by another. But as happens only rarely, this book was handled impeccably by both editors. Thanks to Andrew Miller, now of Random House, for seeing the promise in this project, and to Brendan Cahill, who was a dream editor, the very model of tact, restraint, and understated good taste. A million thanks to Morgan Entrekin, perhaps the last publisher who follows his heart rather than the crowd.

My agent, Kris Dahl, had a faith in this book that exceeded mine. Thanks, Kris, for your persistence, your grit, and most of all, your sense of humor.

Doug Starr, award-winning author and an inspiring teacher, gives new meaning to the word *loyalty*. No one could ask for a better colleague, or a finer friend. Thanks as always for being there, old pal.

As to my little family—Joanna, Alison, and Martin—words aren't enough, but right now they're all I've got. Joanna, thanks for the pile of stuffed bears, the jokes, and the wonderful songs—you are a rock star. Alison, your poise, elegance, insightfulness, and grace sustain and bolster everyone around you. I can't imagine a finer human being. And Marty, whose tolerance, patience, kindness, and deep intelligence I too often take for granted—thanks, Mart, for everything, for every last bit of it.

Index

addictions, 71, 227
adipocytes. *See* fat cells
agriculture, 30–31
  historical perspective on, 24
alcoholic beverages, 32
alkaptonuria (AKU), 52–53
American Dietetic Association (ADA), 198
American Home Products (AHP), 133, 135, 137, 138, 141
American Obesity Association (AOA), 127
Amgen, 102, 113–14, 117, 119
amphetamine, 139
appetite control, 171, 215
appetite gene, 148
appetite regulation, 71–72, 82–84, 122, 165. *See also* leptin; *specific drugs*
  neural pathways that control, 123
Aristophanes, 26
Asia and Asians, 163–64. *See also* *specific countries and cultures*
"atoms of behavior," 86
Atwater, Wilbur Olin, 34–38
Auerbach, Steve, 151, 159
autonomy, 45. *See also* food choices and preferences
back pain, 188
Bahary, Nathan, 92–94, 99, 103
Banting, William, 31–32
bariatric surgery, 12–13. *See also* gastric bypass
Barker, David, 175–76, 180–84
Baywatch, 47–48
behavior modification, 47, 63–64
Bernard, Claude, 31, 32
Billington, Charles, 227
birth weight, 178, 185
Blackburn, George, 13–15, 17, 21
blood pressure, 131–34
Blundell, John, 215
body mass, lean, 148
body mass index (BMI), 3, 88, 155
brain, 61, 84, 123
Bray, George, 130–31, 135, 146
Brown, Steven, 93
brown fat, 148
Bruche, Hilde, 44–45
Buddha, 26
Burger King, 194–95, 199
  Kids Club, 194–95
Burley, Stephen, 100

caffeine, 146
calorie counting, 41, 47

calorie intake, 217
calorimeter, 34
cancer, 75
capitalism, 40
car culture, 222–24
carbohydrate intake, 31, 32, 216. *See also* sugar; sweets
cell proliferation, rapid, 75
children
  advertising and selling to, 191–95, 227
  denied basic autonomy, 45
  overweight and obesity in, 3, 18–19, 45, 185, 186, 227–28
  physical fitness, 187–90
China, 164
Chittenden, Russell, 36–37
cholecystokinin (CCK), 71–76
Christian thinkers, medieval, 27
Chua, Streamson C., 92
Coca-Cola, 212, 214
coconuts, 153, 154
Coleman, Douglas, 50–55, 57–65, 89, 98, 101–4, 106
Coleman, Eric, 131–33
"consumer awareness" groups, 198
consumerism, 40
convenience foods, 201. *See also* fast food industry
"corpulency," 28, 29
Cretans, 26
Crooks, Deborah, 179
"cult of efficiency," 225. *See also* car culture
cultural differences, 3

Darnell, James E., 74, 75, 90
Darwin, Charles, 29
db (diabetic) mice, 58–60, 64, 89, 101
db gene (mutation), 58–59, 89, 90, 92, 94, 119
dexfenfluramine, 134–38
diabetes, 164. *See also db* (diabetic) mice
  pregnancy and, 178, 179
  type 2, 108, 116, 117, 156–57, 160, 178, 185–86
Diamond, Jared, 164
diet
  high-carbohydrate, low-fat, 216
  in Kosrae, 153, 154, 156, 159–60, 169
  low-carbohydrate, high-protein, 31, 32
  poor, 30, 36, 153, 154, 156, 160, 169, 187, 191–92. *See also* fast food
diet books, 47
dieting, "yo-yo," 130

Dionysius, 24–25
diseases associated with overweight, 3–4. *See also* diabetes
DNA, 52, 53
Dole, Vincent, 70–71
Dong, Betty, 127–28
"doubly labeled water," 170
Drewenowski, Adam, 207–8
drug companies, 126–30, 141–42, 147. *See also specific companies*
drug studies, company-sponsored, 127
drug therapy, combination, 149–50
drugs, weight loss, 147. *See also specific drugs*
  and life expectancy, 131
  long-term effects and safety, 131, 143, 147
  producing temporary weight loss, 130, 143, 148
Drury, Ivo, 108
Dutch Hunger Winter babies, 175, 176, 184

Ellis, Michael J., 145
endocrine system, 71–72
endorphins, 71
energy expenditure, 170–71
environmental conditions, 171
environmental cues, 46–47
ephedra, 143–46
ephedrine, 144
Ernsberger, Paul, 136
Escherichia coli (*E. coli*), 100
esophageal banding, 16
evolution, 29
Excerpta Medica, 138
exercise, 170–71, 218–19
externality theory of obesity, 46–47

family dynamics of obese children, 45
Farooqi, Sadaf, 112–15, 117, 121
fast food, 226–27
  heavy users of, 203
fast food industry, 194–209
fat
  and food marketing, 216
  mechanical properties, 50
  physiological functions, 49–50
fat-burning capacity, 218
fat cells (adipocytes), 50
  number of, 83–84
fat intake, 215–16
"fat tax," 232
fat thermostat. *See* "set point"
Federated States of Micronesia (FSM), 154, 155, 159

fen-phen, 133, 134, 137, 140–41
fenfluramine (Pondimin), 133, 134, 137, 138
fertility, 181–82
  body fat and, 83
  Ob mutation and, 115
  starvation and, 165, 174
fetal programming, 176–78
fetal stress, 181–82. *See also* prenatal nutrition
fidgeting, 217, 218
Flemyng, Malcolm, 28–30
Flier, Jeffrey, 108, 117–18, 123, 225
food
  convenience/comfort, 225–26. *See also* fast food
  hedonic impact, 207, 216
  as self-medication against depression/stress, 45
  used to control/pacify child, 45
Food and Drug Administration (FDA), 129, 130, 132, 135, 138
food choices and preferences, 191–92, 198–99, 208–9, 234
food industry, 230–31. *See also* fast food industry
food industry lobby, 198–99, 201–2
food marketing, 191–98, 227
"food stress," 163
Forbes, John, 33
Fosslip, 145
France and the "French paradox," 166–67
Freudian theory, 41–45
Friedman, Jeffrey, 66–71, 73–76, 90–92, 94, 97–99, 103, 112

Galen, 25
Gambia, 165
Garrod, Archibald, 52–53
gastric banding, 16–17
gastric bypass, Roux-en Y, 9–13, 18–22
Gates, Frederick, 69
genetic maps, 90–92
genetics, 28, 85, 88, 120–22, 164–66. *See also* cholecystokinin; Coleman; *db* gene; *ob* gene; O'Rahilly; *specific topics*
  and window of plasticity, 185
genomics, 1. *See also* Coleman; genetics
glucostatic theory, 44
glycerol, 89
Gortmaker, Steven, 190
Great Britain, 181
Greece
  ancient, 24–26
  modern, 184
Greenberg, Isaac, 17–18

Greiser, William, 101–2
guanine, 113

habits. *See* behavior modification; reinforcement
Halaas, Jeffrey, 100, 101
Harvey, William, 28–29
Havel, Peter, 214–16
Hervey, G. R., 61
Hirsch, Jules, 83, 84
homeostasis, 80
hormones, 71–72, 120. *See also specific hormones*
Hummel, Katharine, 58
hunger. *See also* starvation
  inability to recognize true, 46
hypertension, 131–34
  primary pulmonary, 135, 136, 138
hypothalamus, 61, 123

Ibn Sina, 25–26
India, 177
industrialization, 40, 228
insulin, 59–60, 120. *See also* diabetes
insulin resistance, 59
International Obesity Task Force, 127
intragastric balloon, 15–16
ischemic heart disease (IHD), 168

Japan, 213–14
jaw wiring, 15
jejunoileal bypass, 14–15
Jenny Craig, 18
Jonez, Rose, 138–41

Kaplan, Harold, 42
Kaplan, Helen Singer, 42
Kelly, Howard A., 12–13
Kennedy, John F., 189
Keys, Ancel, 80–82, 109–10
Kid Power Exchange, 192
Kirbati, 158
Kluger, Richard, 231
Knoll Pharmaceutical Company, 126–29
Kosrae, 151–57, 159, 166, 168–70, 172
  health problems, 152, 155–57, 159, 162, 163
    obesity, 156, 157
  history, 152–53, 160–62
Kraus, Hans, 188
Kreek, Mary Jeanne, 70
Kuhn, Thomas, 63

Lathrop, Abbie, 56
lean body mass, 148
Leibel, Rudy, 76–79, 82–85, 87–92, 98–99
leptin, 100–102, 112, 113, 116, 118

leptin deficiency, 112, 113
leptin levels, decline in, 116, 117
leptin mutation, 113, 115–16. *See also* Punjabi cousins
leptin pathway, 122
leptin (receptor) gene, 113, 115, 118, 119
leptin resistance, 118, 119
leptin signals and receptors, 118, 119
leptin treatment, 113–17
Levin, Barry, 179
Leyton, G. B., 111
Liebig, Baron Justus von, 30–31
"lipostat," 61–62, 82
Little, Clarence Cook, 56–57
Locke, John, 85–86
Luck, David, 103
Lutwak, Leo, 136–37

Maffei, Margherita, 96
malnutrition. *See* diet, poor; nutrition; starvation
Manson, Joann, 135
Maratos-Flier, Eleftheria, 123
"Margaret," 120
maternal nutrition, 174–80
Maupertuis, Pierre Louis Moreau de, 29
Mayer, Jean, 44
McDonald's, 196–98, 202–4, 206
McNeal, James U., 192, 193
medicine, history of, 69–70
Mediterranean diet, 184
melanin-concentrating hormone (MCH), 123
melanocortin-4 (MC-4), 122–23
Mendel, Gregor, 91
Mendel, Johann, 52
Meridia, 128, 129, 131–33
metabolic syndromes, 109. *See also* Punjabi cousins
Metabolife, 356, 145, 146
metabolism/metabolic rate, 34, 37–38, 148, 149, 164–65, 217
  measuring, 34–35
methamphetamine, 145
mice, laboratory, 55–62
Micronesia, 168, 169, 172
Micronesian One Diet Fits All Today (MODFAT), 159
Mikkelsen, Ejnar, 110
Millennium, 119
Minnoch, Jon, 148
MLN4760, 221
Moller, David, 108
Morgan, Thomas Hunt, 91
mRNA, 96
Mun, Edward, 6–13, 19–21
muscle weakness, 188

narcotic addiction, 71
National Task Force on the Prevention of Obesity, 127
natural selection, 29
Nauru, 157–58
near-starvation. *See* starvation
Neel, James, 162
Neolithic era, 24
neuropeptide Y (NPY), 147–48
"New World Syndrome," 152
non-exercise activity thermogenesis (NEAT), 217
North American Association for the Study of Obesity (NAASO), 125–26
nutrition, 30–31, 36, 78–79. *See also* diet

Ob gene (mutation), 98, 115, 117, 119, 121
ob gene (mutation), 58, 59, 89–92, 94–96, 98–100, 119, 148
ob (obese) mice, 58, 62, 72, 88, 89
ob protein, 89
ob receptor, 100
obesity. *See also* overweight; *specific topics*
  as disease, 2, 28, 29
  historical perspective on, 23–42
  overeating and, 27–28
  psychosocial problems caused by, 18–19
obesity education and prevention, 231–34
"obesity science," 4–5, 121
obesity surgery, 12–18. *See also* gastric bypass
obesity treatments. *See also specific topics*
  ancient, 25–26, 30
ob2j, 96, 98
obsession with food, 111–12
opiate addiction, 71
O'Rahilly, Stephen, 105–9, 111–17, 120–24
oral fixation, 41–42, 44–45
orlistat, 142, 143
overeating, 80
overweight
  degrees of, 33
  prevalence, 3, 160, 168, 230
  as sin or moral failing, 26, 27, 36, 39
  theories of etiology, 30, 37. *See also specific theories*

Paleolithic era, 23–24
pancreas, beta cells of, 120. *See also* diabetes
parabiosis, 60

paradigms, 63
"Patient 24," 115
personality dynamics of obese persons, 41–42
personality traits of obese persons, 45–47
pharmaceutical industry. *See* drug companies
phentermine, 133
physical activity, 170–71, 217–19
physical education (PE), 187–90, 233
physical fitness, 188–89
polymerase chain reaction (PCR), 108
"polysarcia," 30
POMC, 120, 122
Pondimin (fenfluramine), 133, 134, 137, 138
Porter, Philip, 78
prenatal nutrition, 174–85
Prentice, Andrew, 165
primary pulmonary hypertension (PPH), 135, 136, 138
processed food, 205. *See also* fast food
proinsulin, 120
protein, 31
protein intake
  high, 31, 32, 35
  low, 37
psychoanalytic theories of obesity, 41–45
psychosomatic illness, 43
Punjabi cousins, 109–15

Quetelet, Adolphe, 33
Quetelet Index, 33

Ravelli, Anita, 184
Redux, 134–38
Reformation, 28
reinforcement, 47. *See also* behavior modification
religion and overweight, 27, 39
Renaissance, 28
researchers, conflicts of interest, 127
restaurant food, 200, 204, 207. *See also* fast food industry
Rockefeller, John D., 69, 74
rodents, 55–56
Rogers, Oscar H., 39–40
Rubner, Max, 33–34
Russell, Bertrand, 224–25

sacrifices made for being thin, 47
Sasich, Larry, 134
satiation, inability to recognize true, 46
satiety factor, 62, 72, 96
satiety gene, 148

satiety signal, 89, 215
scales, home, 40–41, 47
Schachter, Stanley, 46
Schlosser, Eric, 203
Schneider, Bruce, 71–75, 147
Schoeller, Dale, 170–71
schools, and food industry, 232–33
Schwartz, Hillel, 27
science, 4–5
scientific revolutions, 63
Seigel, Don, 94, 103
self-worth, weight and, 40–41
"set point," 62–63, 82, 221, 222
sexual conflicts, unconscious, 41–42
Sims, Ethan Allen, 79–80, 82, 83
Skilling, Vita, 156
Smith, Steven, 218
social attitudes regarding obesity, 18–19, 27, 38, 40
soda, 211–14
soft drinks, 211–14
Sontag, Susan, 4
starvation, 80–82, 110–11
  and fertility, 165, 174
  during pregnancy, 175
steady state. *See* "set point"
stereotypes associated with overweight, 26–27, 46–47
stomach, 8, 11, 12
stomach balloons, 15–16
stomach shrinking, 16–17
stomach stapling. *See* gastric banding
stomach surgery, 12–15. *See also* gastric bypass
Stunkard, Albert, 43–46, 138
Sturtevant, Alfred, 91
sugar, 210–11, 214, 232
support groups, 18, 19
sweeteners, 210–11, 214
sweets, 213–14
Symonds, Brandeth, 39, 40

Taco Bell, 196
Tartaglia, Louis, 118–19
taste, 209–10
taxing high-fat foods, 232
television watching by children, 186–87, 227
"thrifty fetus" effect, 175
thrifty gene/genotype, 163, 165, 167
thyroid extract, 41
thyroid functioning, 41
tobacco industry, 230–32

ulcers, 13–14
uncoupling protein (UCP), 148–49
U.S. Department of Agriculture (USDA), 234–35

Van Itallie, Theodore, 44
Venus statues, 23–24
vitalism, 87–88
vitamin A deficiency, 159
Voit, Carl von, 31

Wadd, William, 30
weight cycling, 130
weight loss, maintaining, 117
weight loss products and schemes, 4, 18, 126–27, 139
weight regulation, 88
weight stabilization, theory of, 82–83. *See also* "set point"
weight tables, 41
Weight Watchers, 18
Wendy's, 196
Willendorf, Venus of, 23
willpower, 2, 7
Wilson, E. O., 86
Wolff, Harold, 43
World War II, 173–74
Wright, Nancy, 6–11, 19–22
Wyeth-Ayerst Laboratories, 134–38

Xenical, 142–43

Yalow, Roz, 71–74

Zhang, Yiying, 95, 99